As Far as the Eye Could Reach

To Bill,

Wishing you many
happy years ahead,

Your friend,

Phyllis S. Morgan

As Far as the Eye Could Reach

Accounts of Animals along the
Santa Fe Trail, 1821–1880

Phyllis S. Morgan

Foreword by Marc Simmons

Illustrations by Ronald Kil

UNIVERSITY OF OKLAHOMA PRESS : NORMAN

Library of Congress Cataloging-in-Publication Data

Morgan, Phyllis S., 1936–
    As far as the eye could reach : accounts of animals along the Santa Fe Trail.
1821–1880 / Phyllis S. Morgan ; foreword by Marc Simmons ; illustrations by
    Ronald Kil
        pages   cm
    Includes bibliographical references and index.
    ISBN 978-0-8061-4854-0 (pbk. : alk. paper)
1. Animals—Santa Fe National Historic Trail—History—19th century. 2. Santa Fe
National Historic Trail—History—19th century.   I. Title
    QL157.S69M67 2015
    591.978—dc23   2015006781

Copyright © 2015 by the University of Oklahoma Press, Norman, Publishing
Division of the University. Manufactured in the U.S.A.

1 2 3 4 5 6 7 8 9 10

To my son, Thomas M. Morales,

and

to all who work tirelessly and selflessly

to make the animals' lives better

One morning about 9 o'clock, on Turkey Creek, a branch of the Cotton-wood, we came in sight of buffalo, in a great mass, stretching out over the prairie as far as the eye could reach, though the topography of the country enabled us to see for several miles in each direction.

W. B. Napton, *Over the Santa Fé Trail*, 1875

Contents

Foreword

Among the historic trans-Mississippi pathways to the American West, the Santa Fe Trail was the first to be opened, in 1821, and the last to be abandoned, in 1880. It served primarily as a merchant's trail over which flowed a veritable river of U.S. manufactured goods, from the frontier towns of Missouri across the Great Plains to New Mexico's capital at Santa Fe. As an avenue of commerce it stood in sharp contrast to later trails, namely the Oregon Trail and the California Trail, which mainly bore emigrants headed west.

The Santa Fe Trail quickly gained prominence as an important international trade route when Mexico won its independence from Spain in September 1821 and Santa Fe became part of the Republic of Mexico, until 1846 when an American army seized New Mexico at the outbreak of the Mexican War. Thereafter, both ends of the Santa Fe Trail were in U.S. hands.

The international character of the Trail, however, was not entirely lost. As they had done from the beginning, overland merchants upon reaching Santa Fe often continued with their freight wagons down the old Camino Real into the Mexican interior where they sold American goods profitably at national markets and mercantile trade fairs.

Over the long period of its existence, the Santa Fe Trail accumulated a history so rich and varied that, as far as human interest goes, it has no serious rival among the remaining frontier trails that crisscrossed the great West. That fact is reflected in the enormous body of literature dealing with trade and travel on this historic road.

Firsthand accounts in journals, diaries, memoirs, autobiographies and biographies, formal histories, popular narratives, modern travel guides, and even poems have all attempted to tell the tale in their own distinctive way, resulting in a formidable sea of books on the subject. Such sources have been drawn from by other writers to carve out original slices of the Santa Fe Trail and American history that had been overlooked or ignored.

Phyllis S. Morgan in *As Far as the Eye Could Reach* achieves that very thing by sifting through primary and other sources to draw from them eye-witness accounts of wild and domestic animals recorded by Santa Fe Trail travelers. In weaving together the neglected threads of this particular story, she has produced a sum that is far more valuable than the individual parts.

Descriptions of lumbering buffalo or graceful antelopes in flight, for example, when written by assorted observers, begin building in the mind of the reader a truer word-picture of these creatures, their life cycles, and their habits. But most of the men accompanying Santa Fe Trail caravans, as Morgan shows, "caught buffalo or antelope fever" at the first appearance of the big game animals and went rushing off in pursuit of them as newly sighted prey. Any sort of scientific observation was the furthest thing from the minds of these instantly minted hunters.

Not only buffalo and antelope, but all of the other wildlife of the plains elicited abundant comments from men and women who had heretofore known only the natural environment of the eastern woodlands. Prairie dogs and prairie chicken, coyotes and wolves, roadrunners, rattlesnakes, grizzly bears, and mustangs were regarded as consummate curiosities by all new-comers to the Santa Fe Trail.

In addition to the wildlife, the domesticated animals of the traders, travelers, and adventurers taking the Trail greatly increased the number of animals on the land. Domestic animals such as oxen and mules provided the muscle power to pull the heavily laden wagons full of trade goods and supplies, while horses carried individuals over the Trail or drew travelers' carriages, stagecoaches, and other conveyances. And, of course, some people taking the Trail would not leave home without their dogs.

Only from vivid passages in their writing can we today get some idea of the scenes on those vast grasslands. It is a stirring view provided in the pages of Morgan's book, but one touched with melancholy, since her narrative sometimes reminds us of what time and chance have done to the land and its animal life.

Foreword

I am pleased that the author invited me to contribute this foreword, for as a longtime follower of Santa Fe Trail history, I found her *As Far as the Eye Could Reach* to be a worthy addition to the literature of Western trails. I am pleased to extend it my heartiest recommendation.

Marc Simmons
Cerrillos, New Mexico

Preface

*A*s *Far as the Eye Could Reach: Accounts of Animals along the Santa Fe Trail, 1821–1880* consists of essays I wrote over a period of years about the animals seen on the vast, open prairies and plains by the people who followed the Santa Fe Trail, the first and longest-surviving of America's historic trails to the Southwest and West. Preceding the Oregon and California Trails by two decades, this "highway of commerce" originated in Franklin, Missouri, in 1821 and passed through the five present states of Missouri, Kansas, Colorado, Oklahoma, and New Mexico, ending in Santa Fe. Its major routes, the Mountain Route (1,000 miles long) and the Cimarron Route, the shorter and drier route (875 miles long), were heavily used by traders and their trading caravans, travelers, and people seeking adventure until 1880, when the railroad reached the Trail's end. In that nearly sixty-year period, the numbers of wild animals of species that had coexisted on the land with Native peoples for thousands of years decreased significantly. Some species were very close to extinction.

Hundreds of books, essays, articles, and other works have been published about this trail, which was designated the Santa Fe National Historic Trail in 1987. This book is the first to focus on the animals, rather than on the people, places, and events connected to the old Trail. The people who followed its routes during that time in our nation's history are still an important part of my book because they wrote journals, letters, diaries, memoirs, official reports, and books that made the writing of these essays possible. We are fortunate they recorded their observations and experiences, which include lively and colorful descriptions of the different wild animals they saw, many for the first time. Their personal accounts give readers a feeling of being with them on the Trail in the 1800s. From those accounts, I have drawn descriptions of and comments about the wild animals seen in astounding numbers

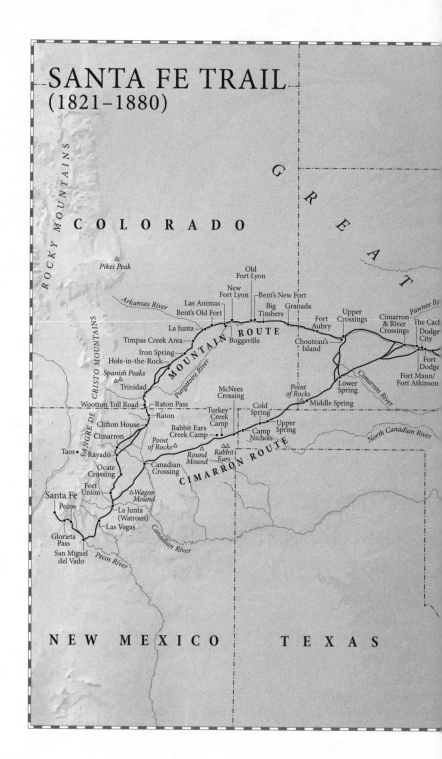

SANTA FE TRAIL
(1821–1880)

ROCKY MOUNTAINS

COLORADO

GREAT

△ Pikes Peak

Arkansas River

SANGRE DE CRISTO MOUNTAINS

Las Animas
Bent's Old Fort
La Junta
Timpas Creek Area
Iron Spring
Hole-in-the-Rock
△ Spanish Peaks
△△ Trinidad

Wootton Toll Road — Raton Pass
Clifton House — Raton
Cimarron
Taos •   Rayado
Ocate Crossing
Fort Union
Santa Fe
Pecos
La Junta (Watrous)
Glorieta Pass   Las Vegas
San Miguel del Vado   Pecos River

Old Fort Lyon
New Fort Lyon — Bent's New Fort
Big Timbers   Granada
MOUNTAIN ROUTE
Boggsville
Chouteau's Island

Fort Aubry
Upper Crossings
Cimarron & River Crossings
Pawnee Ri
The Cach
Dodge City
Fort Dodge
Fort Mann/ Fort Atkinson

Cimarron River

McNees Crossing
Point of Rocks
△ Middle Spring
Lower Spring

Cold Spring
Turkey Creek Camp
Rabbit Ears Creek Camp
Camp Nichols
Upper Spring

North Canadian River

Point of Rocks △
Round Mound
△△ Rabbit Ears

CIMARRON ROUTE

Canadian Crossing

△ Wagon Mound

Canadian River

Purgatoire River

NEW MEXICO   TEXAS

xvi

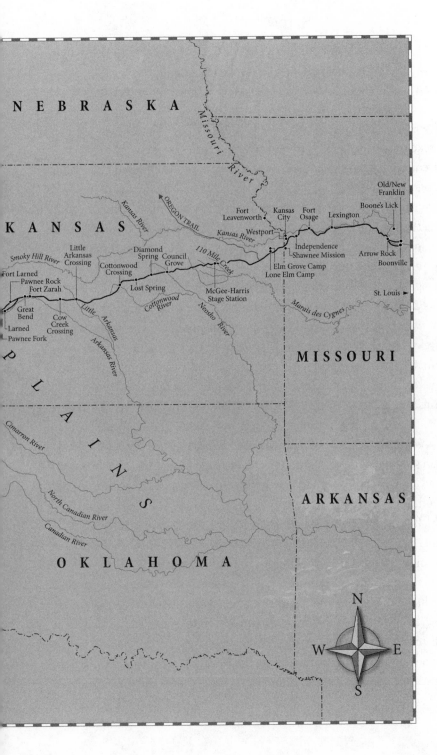

NEBRASKA

Missouri River

KANSAS

Kansas River

OREGON TRAIL.

Fort
Leavenworth

Kansas City

Fort
Osage

Lexington

Boone's Lick

Old/New
Franklin

Kansas River

Westport

Little
Arkansas
Crossing

Diamond
Spring

Cottonwood
Crossing

Council
Grove

110 Mile Creek

Independence

Shawnee Mission

Elm Grove Camp
Lone Elm Camp

Arrow Rock

Boonville

Smoky Hill River

Fort Larned
Pawnee Rock
Fort Zarah

Lost Spring

McGee-Harris
Stage Station

*Cottonwood
River*

Neosho River

Marais des Cygnes

St. Louis ►

Great
Bend

Larned
Pawnee Fork

Cow
Creek
Crossing

Little Arkansas

Arkansas River

MISSOURI

P   L   A   I   N   S

Cimarron River

ARKANSAS

North Canadian River

Canadian River

OKLAHOMA

N

W   E

S

xvii

on the land and about the domestic animals that belonged to the people who took the Trail or were part of the caravans.

Most astonishing to the travelers were the incredible numbers of animals seen and encountered along their way over the Santa Fe Trail. In their writings, they provided descriptions for families, friends, and others so that those readers might be able to visualize the scenes travelers were seeing. They frequently spoke of animals "as far as our eyes could behold," "as far as the eye could reach," and "in all directions beyond the reach of the eye." The title of my book was taken from one of these phrases. One can only imagine the stir of emotions and feelings those travelers experienced, from delight to utter amazement to chilling terror, especially the first-timers who were venturing into the unknown.

During the extensive research, reading, and gathering of travelers' accounts and anecdotes about the animals, in particular the wild animals, it became evident which of those should be included in my book. Time and again, the travelers wrote about these animals, which were new and unusual to them and aroused their awe, curiosity, excitement, and wonderment. Although there were other wild animals present, especially through the region of the Arkansas River, the travelers seldom wrote about those they recognized or knew back home, such as rabbits, deer, turkeys, and geese. These familiar animals were sometimes briefly mentioned when hunters brought animals they had shot back to the camp for food. Travelers spent the little free time they had to write about their new experiences. Also, they seldom wrote about animals seen more frequently in other regions of the plains, such as elk, which were plentiful along the Canadian River region farther south of the Santa Fe Trail, or grizzly bears, which were seen more frequently on the Mountain Route or in the foothills of the Rocky Mountains.

Expanding upon the Santa Fe Trail travelers' and traders' accounts, I have included many facts and interesting information about the animals. They have incredible abilities and are fascinating as individuals and as members of their respective families or species. Revealing anecdotes also tell about the human-animal relationships along the Trail. I hope this book will help readers think about wild and domestic animals, those living in the past and today, as more than just "dumb animals," a most undeserving epithet.

I believe, after extensive reading and research about the Santa Fe Trail, the frontier, and the opening of the American West, that animals have been given short shrift when it comes to their rightful place in the history of our country. They are often overlooked or given little attention, though millions of animals have played an important role in American history. Usually, they have not been considered to have a place in our national story. I believe it is time to acknowledge them and their efforts and suffering in building our great nation. This is my most important reason for writing these essays and bringing them together in this book.

Many memorable experiences have helped me greatly in writing these essays. The idea of writing them came to me as I stood in the ruts and swales made by the wheels of those large, heavily laden wagons that once crossed the land. Those remnants of this historic "highway of commerce" can still be seen in many places to this day. My curiosity and interest in the Trail brought me to the association that bears its name, the national Santa Fe Trail Association. The association and its chapters, located across the old Trail from Missouri to New Mexico, have taken the lead since the founding of the association in 1986 in commemorative and historical endeavors to keep the Trail alive. I began reading all types of materials, studied maps, and attended symposiums held by the association. Shortly after joining it, I learned that a small group of hikers, led by author Inez Ross of Los Alamos, was hiking the Trail from Santa Fe to mid-Missouri, where it first began, and I gladly joined them. That hike lasted ninety-seven days over a period of years on the Cimarron Route, also known as the Cimarron Cutoff or Dry Route. The 875 miles were completed in the summer of 2004. This marvelous experience of retracing history and being on the land will remain with me until the end of my days, just as it did for many of the early Trail travelers.

In addition to experiencing the Trail on foot, I have driven along its routes countless times, stopping to see familiar places and new wayside exhibits and other recent additions providing new information and insights about the Trail and its history. I have also flown over it a few times, in particular when I accompanied men and women who fly small private planes over the historic trails of the United States and especially of the American West. They call themselves "The Historic Trail Flyers." They head out every

September to experience a historic trail's route from the air and then on the ground with knowledgeable guides. It is amazing what still can be seen on the ground and from the air, especially when one considers that 2021 will mark the two hundredth anniversary of the opening of the Santa Fe Trail.

Phyllis S. Morgan
Albuquerque, New Mexico

Acknowledgments

Most of the essays in this book were previously published over a number of years in *Wagon Tracks*, the quarterly journal of the Santa Fe Trail Association. I thank the association and the former editor (now retired), historian Leo E. Oliva, for the publication of these writings. Dr. Oliva also deserves my deepest gratitude for sharing his knowledge about various aspects of the Trail's history. The essay about rattlesnakes on the Trail also appeared in an issue of *Reptiles,* and I thank the editor for his kind permission to reprint it. The essays about the animals have been revised and added to for this book. Excerpts quoted from the travelers' accounts appear in this book as they appear in the sources from which they were drawn. Words or phrases enclosed in brackets have been inserted for clarification or to define terms.

My sincere thanks are extended to all at the University of Oklahoma Press for their efforts pertaining to my book. In particular, I extend sincere gratitude to acquisitions editor Alessandra Jacobi Tamulevich, editorial assistant Thomas Krause, managing editor Steven Baker, and copyeditor Norma McLemore. My heartfelt appreciation also goes to Marc Simmons, recognized authority on the Santa Fe Trail, for his foreword and to artist Ronald Kil for his marvelous animal illustrations and cover art for my book.

In researching this book, which began in earnest in 2001, I am indebted to countless librarians and other staff of academic, public, and private libraries, and to those of the historical societies in the states through which the old Santa Fe Trail crossed. Special thanks to the Center for Southwest Research and Zimmerman Library at the University of New Mexico in Albuquerque and to the Special Collections Department of the Albuquerque–Bernalillo County Library System.

Acknowledgments

Many individuals were also very helpful in providing assistance and much appreciated encouragement, especially Donald C. Dickinson, Rick and Sharon Hannen, Donna Nelson, Susan Phelps, LeRoy Anthony Reaza, James Weaver, Wyman Meinzer, Bob Meyers, Valerius Geist, and the late Katharine B. Kelley.

Also, a special acknowledgment goes to the officers, board of directors, and members of the Santa Fe Trail Association for preserving and protecting the Trail, educating all ages about it, and marking its historic routes exceptionally well so that present-day travelers can follow the Trail and learn about an important period in our nation's history.

Part I

Wild Animals on the Santa Fe Trail

Chapter One

Buffalo

Buffalo were the most important wild animals on the prairies to the travelers on the Santa Fe Trail. They were the focus of attention when the caravans reached buffalo country. Without them to provide the life-sustaining protein and other nutrients required for the arduous journey from Missouri to New Mexico and back again, the traders and others on the Trail, as well as the early settlers on the plains, would most likely not have fared as well as they did. Indians had depended on these animals for thousands of years, and the Spanish colonists and *ciboleros* (buffalo hunters), relied on them for more than two hundred years before the trade caravans arrived in the 1820s.

During his first journey over the prairies to Santa Fe, merchant William Becknell, remembered as "the Father of the Santa Fe Trail," recorded in his journal on September 24, 1821: "We reached the Arkansas [River], having traveled during the day in sight of buffaloe, which are here innumerable."[1]

During the surveying of the Santa Fe Trail, George Champlin Sibley, who led the survey conducted from 1825 to 1827, reported: "The Road, in nearly its whole extent passes over open, grassy prairie. . . . Caravans may obtain their chief Supplies for Subsistence, without difficulty or delay, from the numerous herds of Buffaloes that are almost continually passing and repassing over the plain, crossing the Route everywhere along the greater part of the way; and many years must elapse before this great Resource will fail, or materially diminish."[2]

In those days, herds of "the monarch of the plains" made a magnificent scene to behold, each animal an impressive member of the animal kingdom. The largest land mammal of North America, a male buffalo, or bull, can reach a height of six to six-and-one-half feet at the shoulders, seven to eleven feet in length, and weigh nearly a ton. Once ranging as far east as New England

and the Atlantic coast, these massively built animals reigned supreme over the prairies and woodlands for thousands of years. Roaming is in their nature; about three hundred thousand years ago they roamed across the Bering land bridge from their home in Asia to North America.

This member of the bovid family, Bovidae, is not a true buffalo, but a species of bison. Called the American bison (*Bison bison*), this bovid is distinguished from the true buffalo, such as the Cape buffalo of Africa and the Asian water buffalo, by the hump on his shoulders and his extra pair of ribs. The American bison has fourteen pairs of ribs, while the true buffalo has thirteen pairs. The early French explorers called this animal *bœuf,* meaning ox, changing over time to buffalo. The Spanish word for buffalo is *cíbola.*

A number of travelers on the Santa Fe Trail were aware that "bison" is the precise and scientifically correct term, but still preferred "buffalo," the name that continues to be more popular. In a country where words are frequently replaced because the former belonged to another time, no one has reported hearing the coin that pictures the distinctive profile of this American icon referred to as a bison nickel. Nor has anyone insisted on changing "buffalo" to "bison" in Kansas pioneer Dr. Brewster M. Higley's well-loved "The Western Home," written in 1873 and known to all as "Home on the Range." It would be unthinkable for William F. Cody fans to start referring to him as "Bison Bill." The word "buffalo" has made a deep and lasting impression on the American psyche, and it seems certain that "buffalo" and "bison" will be used interchangeably for a long time to come.

Colonel Richard I. Dodge, who commanded Fort Dodge, Kansas, from 1872 to 1873, explained why he chose to use "buffalo" in his long chapter about this gregarious animal in *The Plains of the Great West and Their Inhabitants* (1877): "I suppose I ought to call this animal the 'bison'; but, though naturalists may insist that 'bison' is his true name, I, as a plainsman, also insist that his name is buffalo. As buffalo he is known everywhere, not only on the plains but throughout the sporting world; as buffalo he lives and moves and has his being; as buffalo he will die; and when, as must soon happen, his race has vanished from earth, as buffalo he will live in tradition and story."[3] Dodge's last words here are foreboding. In the 1870s, buffalo were already in serious decline.

Colonel Dodge, Indian leaders, trader and chronicler of the Santa Fe trade Josiah Gregg, artist George C. Catlin, and others who knew the Trail foresaw the grim future of the buffalo and their demise on the plains. Some expressed their beliefs that this important animal was in danger of extermination. Most travelers on the plains, however, paid no heed, positive that the buffalo would last forever.

People following the routes of the Santa Fe Trail, the Mountain Route and the Cimarron Route, sometimes referred to as the Cimarron Cutoff or Dry Route, seldom overlooked describing or commenting on the sight of buffalo herds. They had heard many stories and read about them in books and newspapers. They imagined what buffalo country would be like, but they were totally astounded by what they saw, a scene far beyond their imaginings or expectations.

Lydia Spencer Lane saw buffalo on a number of occasions in travels with her army husband. She wrote about them during a trip over the Trail in the 1850s from Fort Union in New Mexico to Kansas City, Missouri, traveling with a party of ten-mule wagons for twenty-four days. She also told of the many perils of hunting on the land where the buffalo roamed: "In those days the whole country was covered with immense herds of buffalo; there were thousands and thousands of them; yes, a million. They never molested the trains crossing the Plains, though sometimes a great drove of them came thundering down to the road, and the wagons were obliged to halt until they passed.

"There was no difficulty in killing one when fresh meat was needed; but the wary hunter seldom wandered far away, as there were plenty of Indians abroad as well as buffalo. A man strayed off one day, and we knew nothing of him until night, when he came into camp, naked. Indians had caught him while hunting, taken all his clothes, even his shoes, and then turned him adrift. He kept at a respectful distance from the wagons until darkness covered him, the only mantle he had, and then came into camp. He did not care much for hunting during the rest of his travels."[4]

In his popular book *The Old Santa Fe Trail* (1939), Stanley Vestal described the heart of buffalo country: "Far and wide, on every hand, the sign of those majestic animals was to be seen, and at all seasons. . . . Everywhere the soil had been scooped into shallow, saucer-like depressions by wallowing bison. These wallows were indestructible, unmistakable from their circular shape—though they varied in size from four or five to fifty feet across. . . . In any season, those wallows were an unfailing sign that buffalo ranged the country. And now, as gray wolves were seen insolently trotting along the ridges, everyone knew the herds could not be far off. Every man in the caravan felt his blood begin to heat with buffalo fever."[5]

This feverish affliction raged across the prairies and plains. It gripped the men, young and old, as they drew closer to the far-flung region on both sides of the Arkansas River. Their expectations of shooting and killing buffalo became palpable. In the excitement, the chase would sometimes turn into riotous confusion with buffalo and hunters going in every direction, creating a dangerous situation for all. Many buffalo were killed only for sport. Historian Marc Simmons has commented: "It was a kind of primitive blood lust that led to pointless slaying of buffalo."[6]

Philip Gooch Ferguson, a company clerk of the First Regiment of the Missouri Mounted Volunteers in General Stephen Watts Kearny's Army of the West during the Mexican War, caught "buffalo fever" as he headed over the Trail to Santa Fe. He recorded in his diary during a rest stop on July 22, 1847: "We came in sight of black-looking masses on the prairie, which some said were buffaloes, but others could scarcely believe it. There were such numbers of them. Yet buffaloes they proved to be, and forgetting our duty as scouts, we determined to give them a chase. Lindemore [another regiment member] and I rode out to drive them down, but this proved a vain attempt, for as soon as the herd took the alarm, they broke off and could not be headed.

"Finding we could not turn the course of this living current, I determined to have a shot at them as they passed. But Black Hawk [Ferguson's horse], not fancying the looks of the shaggy animals, refused to go close to them, and I fired my musket at an old bull over a hundred yards distant while my horse was at full speed. But if I hit him, he did not mind it, as he continued on at a rolling gallop. . . . Being nearly exhausted myself and having run my horse about six miles, I returned to the company, which was now out of sight. Thus ended my first buffalo chase, and I had fully experienced the wild excitement it inspires."[7]

William B. Napton, Jr., was eighteen in June 1857 when he left the West-port, Missouri, area (close to Kansas City) in a train of twenty-six wagons headed for Santa Fe. Two of the wagons were loaded with bottles of champagne for Colonel Céran St. Vrain, former mountain man and partner of Charles and William Bent, whose Bent's Fort on the Mountain Route was a renowned landmark. That year about 12,000 wagons left the Kansas City area for Santa Fe; 9,884 of those went to New Mexico.[8]

The son of a wealthy family residing in Saline County, Missouri, young Napton was well educated, proficient with horses, and skilled in the use of guns. His health, however, was "indifferent," and his father thought a trip over the Trail would help to improve it. He was able to acquire a well-trained, experienced buffalo horse and looked forward to his first buffalo chase.

Napton did not have to wait long. His first chase proved exhilarating but unsuccessful. He later wrote about his initial attempt at hunting buffalo: "As we were drawing near the buffalo range, preparations were made for a chase. The pistols were freshly loaded and butcher knives sharpened. . . . One morning about 9 o'clock on Turkey Creek, a branch of the Cotton-wood, we came in sight of buffalo, in a great mass, stretching out over the prairie as far as the eye could reach, though the topography of the country enabled us to see for several miles in each direction.

"We rode slowly until we got within three or four hundred yards of the edge of the vast herd. Then they began to run and we followed, gaining on them all the time. Pressing forward, at full speed of my horse, I discovered that the whole band just in front of me were old bulls. I was so anxious to kill a buffalo that I began shooting at a very large one, occasionally knocking tufts of hair off his coat, but apparently having little other effect. However, after a lively run of perhaps a mile or two he slackened his pace, and at last stopped still and, turning about, faced me. I fired the one or two remaining charges of my revolver, at a distance of twenty or thirty yards, and thought he gave evidence of being mortally wounded."[9]

The old buffalo looked intently and steadily at Napton for a few minutes, then turned and walked away. Napton followed the buffalo until the animal began galloping toward the main herd and disappeared behind a ridge. Disappointed over his failure and worn out by the chase, Napton headed back to the wagon train. The captain of the caravan laughed at him at first but gave the young man hope for a better second attempt. Instructed by the captain "as to the *modus operandi* of killing buffalo on horseback at full speed," the young adventurer resolved to try another chase.

The next morning, he mounted his rested, eager steed and "sallied forth." The buffalo still filled the scene in front of him. Napton wrote in his reminiscences: "At the left of the road, in sight, thousands of buffalo were grazing in a vast plain, lower than the ridge down which we were riding. Opened up in our view was a scope of country to the southeast of us, a distance of ten miles. This plain was covered with them, all heading towards the northwest."[10] Obviously a quick learner, he rode his horse up next to a fat cow and, with his second shot, brought down his first buffalo.

The young hunter was faced, however, with a difficult situation. His shots had set the herd in motion, and he was completely surrounded by the rushing mass of buffalo. "The air," he recalled, "was so clouded with dust that I could hardly see more than twenty yards from where I was standing, near the carcass of the cow I had killed. There was danger of being run over by them, but they separated as they approached, passing on either side of me, a few yards distant. After a while the rushing crowd thinned."[11] The captain had been watching the action and rode up close to Napton. He urged the young man to try to shoot another buffalo. Napton headed his horse into the midst of the herd and was able to kill one more buffalo. He had managed to kill two buffalo in less than half an hour. After the second chase, he did not experience any problems in killing all of the buffalo his company needed for food.

Napton marveled at the great number of buffalo and at the quality of buffalo meat: "For a week or ten days they were hardly out of sight. We found them as far west as Pawnee Rock. All told, I killed about twenty on the journey out and back. A good steak, cut from the loin of a buffalo cow, broiled on the coals with a thin slice of bacon attached to it to improve its flavor, was 'good eating,' and I soon became an accomplished broiler."[12]

All agreed that the numbers of buffalo were "immense" and "overwhelming." Before the opening of the Trail in 1821, Zebulon Montgomery Pike recorded in a journal of his 1806–1807 expedition through the Louisiana Purchase territory: "I will not attempt to describe the drove of animals we now saw on our route [in Kansas headed west of present-day Cimarron];

suffice it to say that the face of the prairie was covered with them, on each side of the river; their number exceeded imagination."[13]

The day before, Pike had climbed a hill to watch the action as members of his party took a break to kill some buffalo cows and calves for their food supply. He wrote that the scene "gave a lively representation of an engagement. The herd, having been divided into separate bands, first charged on one side and then to the other, as the pursuit of the men on horseback impelled them. . . . The report and smoke from the guns, added to the pleasure of the scene, which in part compensated for our detention. The cow buffalo was equal to any meat I ever saw, and we feasted sumptuously on the choice morsels."[14] Among the morsels considered delicacies were tongue, liver, marrow, and the fat anterior portions of the buffalo's hump.

Like Pike and Napton, most of the Trail travelers considered buffalo meat, except that from tough, old bulls, to be more savory and juicier than beef, probably because the fat is more evenly distributed in the meat. Frank S. Edwards, a Missouri Mounted Volunteer in the Mexican War, had a differing opinion, although his first impression may have been affected by the age of the buffalo and the way it was cooked. He wrote in camp at Pawnee Fork on July 15, 1846: "Here I first tasted buffalo meat. Our hunters, who were selected from the companies each morning, had been successful in killing three out of an immense herd which we had seen crossing a roll of the prairies during the day. There must have been over three or four thousand in the herd, and from the distance, they resembled a shadow cast upon the earth from a black cloud as it passes across the sun. The buffaloes killed consisted of two old tough bulls and a nice young cow, the latter of which Antoine, our hunter had taken. . . .

"On account of the entire absence of wood here, we had to use the dry dung of the buffalo, called by the hunters *bois de vâche* [cow wood] or buffalo chips, for fuel. There was plenty of it around our camp, and it had one advantage over wood—it required no chopping. It makes a good and hot fire without flame, but had a strong ammoniacal odor, which is imparted to everything cooked by it. Our buffalo meat, which we simply roasted on the live embers, of course partook largely of this flavor. . . . To tell the truth, I was much disappointed in the flavor of buffalo meat, and would rather have a piece of good beef."[15]

Edwards also commented on the buffalo wallows: "The mud-holes where they roll or wallow, become, sometimes of very large size, from these living mud-scows carrying off, one after another, considerable quantities of the moist soil. . . . The rain forms them into ponds, and fish are frequently found in them."[16] Wondering about such a phenomenon as fish in a buffalo wallow in the midst of the plains, Edwards ended his entry: "Where do these fish come from?" Whether Edwards found the answer to that question, we will never know.

The buffalo were said to number sixty million, a figure that old-timers stuck to religiously. Dale F. Lott, wildlife biologist, remarks in his book *American Bison: A Natural History* (2002): "'Sixty million bison,' has long been as close to religious dogma as a secular society's beliefs can be. . . . Such importance justifies a really close look at how that figure got fixed in our collective consciousness."[17]

Lott tells how scientist Jim Shaw traced this number back to the works of Ernest Thompson Seton, naturalist and author of many popular books about wild animals. Seton based his calculations on Colonel Dodge's account of the herds seen during a thirty-four-mile trip made in May 1871 in a light wagon over the Santa Fe Trail in Kansas from old Fort Zarah west to Fort Larned.

Dodge described that trip as follows: "At least twenty-five miles of this distance was through one immense herd, composed of countless smaller herds, of buffalo then on their journey north. The road ran along the broad level 'bottom,' or valley, of the river. Some few miles from Zara [Fort Zarah] a low line of hills rise from the plain on the right, gradually increasing in height and approaching the road and river, until they culminate in Pawnee Rock when they again recede. The whole country appeared one mass of buffalo, moving slowly to the northward; and it was only when actually among them that it could be ascertained that the apparently solid mass was an agglomeration of innumerable small herds, of from fifty to two hundred animals, separated from the surrounding herd by greater or less space, but still separated. . . .

"When I had reached a point where the hills were no longer more than a mile from the road, the buffalo on the hills, seeing an unusual object in their rear, turned, stared an instant, then started at full speed directly towards me, stampeding and bringing with them the numberless herds through which they passed, and pouring down upon me all the herds, no longer separated, but one immense compact mass of plunging animals, mad with fright, and as irresistible as an avalanche. The situation was by no means pleasant."[18] Dodge reined up his trusted steed, an old buffalo horse accustomed, fortunately, to stampeding buffalo.

"I waited until the front of the mass was within fifty yards, when a few well-directed shots from my rifle split the herd, and set it pouring off in two streams, to my right and left. When all had passed me, they stopped, apparently perfectly satisfied, though thousands were yet within reach of my rifle, and many within less than one hundred yards. Disdaining to fire again, I sent my servant to cut out the tongues of the fallen. This occurred so frequently within the next ten miles, that when I arrived at Fort Larned I had twenty-six tongues [a prime delicacy] in my wagon, representing the greatest number of buffalo that my conscience can reproach me for having murdered on any single day. I was not hunting, wanted no meat, and would not voluntarily have fired at these herds. I killed only in self-preservation, and fired almost every shot from the wagon."[19]

According to Dale Lott, Seton had a keen mind for calculations. He tried every reasonable way to estimate the population of buffalo before the Great Slaughter. Since the time Seton made his decision to stick with sixty million, scientists have used a variety of approaches, including studying the land's carrying capacity, to come up with a figure. Their attempts may be more sophisticated, but the result is still speculation. Today, some biologists think the bison population might have been closer to thirty million.[20] Lott prefers to use "tens of millions" to describe the number of buffalo once roaming the prairies and plains.

Some Trail travelers recorded their thoughts about the physical appearance and behavior of buffalo. The early Spanish explorers considered them the most monstrous-looking animals ever seen. Much later, Kentuckian Susan Shelby Magoffin, wife of Santa Fe Trail trader Samuel Magoffin, jotted in her diary while resting at Big Coon Creek on July 13, 1846: "Passed a great many buffalo (some thousands), they crossed our road frequently within two or three hundred yards. They are very ugly, ill-shapen things with their long shaggy hair over their heads, and the great hump on their backs, and they look so droll running. . . . They draw themselves into a perfect knot switching their tails about, and throwing all feet up at once."[21]

Josiah Gregg noted in his classic book *Commerce of the Prairies* (1844): "It has been the remark of travelers that the buffalo jumps up from the ground differently from any other animal. The horse rises upon his fore feet first, and the cow upon her hind feet, but the buffalo seems to spring up on them all at once."[22] Albert Pike, a young Bostonian who followed the Trail on his way to Taos, described them as "heavy, unwieldy" animals that "seem, even at their best speed, to be moved by some kind of clumsy machinery."[23]

Although their enormous bulk, humps, large heads, long front legs, and skinny rumps may make them look awkward to the casual observer, buffalo have impressive athletic abilities. Their long front legs give them a remarkable stride to escape the fangs of their animal predators. They can swiftly pivot and turn around, gallop at more than thirty miles per hour, and leap tall road cuts in a single bound. Dale Lott has observed them over many years at the National Bison Range in western Montana: "They are capable, at any second, of a memorable athletic moment."[24] He watched a two-thousand-pound bull do a standing high jump of six feet. Buffalo can also perform standing broad jumps of fourteen feet (their jumps are described as "hops"), which is why the National Bison Range has extended cattle guards.

By the mid-1800s, Indians and buffalo were seen as impediments to the westward movement of settlers and to the conversion of plains into range for cattle and tillable farmland. In an unspoken national policy, the federal government formed its plan to eradicate the buffalo and remove the Indians to reservations. In 1873, Columbus Delano, then U.S. secretary of the Interior, wrote: "The civilization of the Indian is impossible while the buffalo remain on the plains."[25] Hide hunters, supplied with free ammunition and protected by the U.S. Army, became the instrument of that policy beginning in 1868. The hunters began shooting the buffalo in earnest in the early 1870s; it took until 1883, when the last large hunt occurred, to eliminate nearly every one of them.

Millions of hides and tongues, and tons of meat, were sent to eastern destinations, although countless carcasses were left to rot on the land. The last commercial shipment of hides was in 1889. The bone collectors followed, gathering the bleached skeletons and bones littering the plains. The Great

Slaughter had ended, and the buffalo were gone from the land and the routes of the old Santa Fe Trail, which had already passed into history in 1880 with the arrival of the railroad in Santa Fe.

By 1894, the only free-living buffalo remaining in the United States were found in Yellowstone National Park. By 1902, this last wild herd had been reduced by poachers to twenty-three survivors. A few hundred other buffalo survived on private land. In an ironic turn of events two decades later, the U.S. Army, which managed Yellowstone National Park, played a role in the buffalo's return from near extinction.

These remarkable animals, revered and honored with the name "Uncle," had supplied Indians with nearly 100 percent of the raw materials they needed to survive. A marker erected by the Kansas Historical Society near the old Santa Fe Trail in southwestern Kansas tells how much they depended upon the buffalo: "The buffalo was the department store of the Plains Indian. The flesh was food, the blood was drink, skins furnished wigwams, robes made blankets and beds, dressed hides supplied moccasins and clothing, hair was twisted into ropes, rawhide bound tools to handles, green hides made pots for cooking over buffalo-chip fires, hides from bulls' necks made shields that would turn arrows, ribs were runners for dog-drawn sleds, small bones were awls and needles, from hooves came glue for feathering arrows, from sinews came thread and bow-strings, from horns came bowls, cups and spoons, and even from gall stones a 'medicine' paint was made." More uses could be added to this list; for instance, the tail was useful for whisking away flies.

Without the buffalo, the Indians were destitute. In 1882, only a decade after he saw the herd that covered the prairie for twenty-five miles, Colonel Dodge declared: "Ten years ago, the Plains Indians had an ample supply of food.... Now, everything is gone, and they are reduced to the condition of paupers, without food, shelter, clothing, or any of the necessaries of life which came from the buffalo."[26]

Mike Fox, a director of the Fish and Game Department of a northern Plains Indian tribe and the person responsible for the reservation's buffalo herd, told Ruth Rudner, author of *A Chorus of Buffalo* (2004): "It's kind of full circle. Right now, it's our turn to take care of the buffalo. In the very near future, they'll be taking care of us again. In the past, you know, they totally took care of us." He added: "Humans have to be part of the management. In the old days, we didn't call it management. We called it survival. We'd take X number of animals. Dealing with these animals, you know that at some point you're going to have a surplus if you don't use them as they're meant to be used."[27]

In recent years, buffalo have been gaining a hoofhold in a number of areas where the herds once roamed in huge numbers. Most live in national and state parks, wildlife refuges, and on private land, including the large buffalo ranches owned in several states by Ted Turner. In Kansas, where buffalo are the state animal, they live near the Mountain Route of the old Santa Fe Trail at the Finney Game Refuge, outside of Garden City. The oldest publicly owned buffalo herd in the state lives at this refuge.

Although their numbers are increasing, the future for buffalo is far from secure, particularly in places where people see potential problems with their return. One concerns the competition between buffalo and cattle for food and water. Another is the fear that buffalo are carriers of brucellosis, a bacterial disease that may endanger cattle. Buffalo and cattle, however, generally do not graze together, and no credible scientific study has proved transmission of brucellosis from buffalo to domestic livestock under natural pasture conditions.[28] Vaccines are available to prevent brucellosis in cattle and buffalo, but these are not 100 percent effective.

Today, it is a thrilling sight to behold a herd of buffalo, although the number may be minuscule in comparison to the vast herds seen by the people on the Santa Fe Trail, particularly during "the Golden Age of the Trail," from the 1820s to the early 1860s. The lives of buffalo depend on the advocacy and dedication of concerned people from all walks of life who want this majestic American icon to survive and thrive in their natural habitat.

Pronghorn

Pronghorn (*Antilocapra americana*) are the fastest mammal in the Western Hemisphere and among the fastest in the world. They were observed and commented on by European explorers and other early travelers to North America. The Spanish named them *berrendo,* and the French Canadians called them *cabrie.* Indians on the plains, where millions of pronghorn once roamed, gave them many different names; twenty have been recorded.[1]

The name "antelope" was given to the pronghorn by Captain Meriwether Lewis during the Lewis and Clark Expedition of 1804–1806. The Corps of Discovery saw their first pronghorn on September 3, 1804, and Captain Lewis was the first to record this animal and collect a specimen for science.[2] Thus, "antelope" was the name generally used by English-speaking travelers on the Santa Fe Trail and has been used colloquially ever since. Some people who saw them, including William Clark, Lewis's co-captain, called these animals "goats."

Today, however, the accepted common name is "pronghorn," derived from the short prong on their horns, which distinguishes them from all other animals. Although reference is occasionally made to "pronghorn antelope," the name "pronghorn" is used by those who study and write about this unique, fascinating animal. In this chapter, "antelope" is used when the people who followed the Santa Fe Trail used it in their writings or when early authors used it consistently in their works.

Many travelers on the Santa Fe Trail who kept journals, personal accounts, or official reports noted a great abundance of antelope. Some only mentioned seeing them during the day's journey, while others added the number counted, whether they were shot at, and if the shots missed or killed them. Others provided more detailed descriptions of these beautiful, swift animals

and their characteristics, in particular their amazing fleetness and their intense curiosity and penchant for checking out any unusual activities.

Augustus Storrs, who served as the U.S. consul in Santa Fe, first journeyed over the Trail from Missouri to Santa Fe in 1824, three years after William Becknell opened the Trail to international commerce. Later that year, Storrs wrote one of the first reports about it published by the U.S. Senate. He had been asked by Thomas Hart Benton, U.S. senator from Missouri and a strong trade advocate, to submit a report in support of legislation for an official survey of "the Road," referring to the Santa Fe Trail. Storrs wrote in his report published in January 1825: "With regard to the natural means of subsistence, there is probably no other equal extent of wilderness in the world so well supplied. Deer are scarce, but buffalo, elk, and antelope are abundant. . . . Our company had an ample supply of fresh meat almost every day."[3] Congress quickly passed legislation establishing a three-man commission to survey and mark the Trail. This survey, conducted from 1825 to 1827, was led by George Champlin Sibley, who had considerable experience on the frontier as the factor, or government trader, at Fort Osage in Missouri. This fort had been established by William Clark, co-captain of the Corps of Discovery, in 1808 and for a brief time served as a rendezvous for caravans traveling the Trail.

Another early traveler, Alphonso Wetmore, was an army paymaster stationed in Franklin, Missouri, at the eastern end of the Trail. He made notes for a report, including the abundance of pronghorn and wrote on June 19, 1828: "The antelope is a subject of speculation this morning, and one of our hunters has been occupied in decoying, with a flag, one of these nimble-footed animals."[4] Wetmore was referring to "toling" game, which entails a variety of methods used by hunters to lure or decoy game by arousing curiosity, so that the animal would approach within range of their guns. A number of men mentioned toling in their writings.

Youthful adventurer Lewis H. Garrard wrote about the antelope on his trip across the prairie in 1846. A keen observer, he described this colorful animal in his book *Wah-to-yah and the Taos Trail* (1850): "There is much that is singular about the antelope, it being a most inquisitive creature; their curiosity, like Eve's, often results in their downfall. While hunting before reaching camp, a band came running past. Bewildered & fascinated, they described two complete circles around us, during which we gave them several shots, though their motions were too swift for sure aim."[5]

Garrard also described a popular way of toling antelope, which entailed a hunter standing on his head and shaking his legs in the air. An expert at this type of toling, according to Garrard, was Marcellin St. Vrain, the leader of Garrard's group and the brother of Céran St. Vrain, partner of William and Charles Bent. He also mentioned: "A handkerchief on a gun rod will cause them, now advancing, now retreating, to approach until within rifle range."[6]

Trader Josiah Gregg saw countless antelope during his four round trips on the Trail with trade caravans from 1831 to 1840. He wrote about the fascinating antelope and toling them in *Commerce of the Prairies* (1844): "Being as wild as fleet, the hunting of them is very difficult, except they be entrapped by their curiosity. Meeting a stranger, they seem loth to leave him until they have fully found him out. They will often take a circuit around the object of their curiosity, usually approaching nearer and nearer until within rifle-shot, frequently stopping to gaze. Also, they are often decoyed with a scarlet coat or a red handkerchief attached to the tip of a ramrod, which will sometimes allure them within reach of the hunter's aim. But this interesting animal, like the buffalo, is now very rarely seen within less than two hundred miles of the frontier: though early voyagers tell us that it once frequented as far east as the Mississippi."[7]

Gregg was interested in natural history and science, and kept many records of his observations along the Trail. In describing the antelope, he incorrectly identified these animals as gazelles. (The pronghorn of North America are not related to the true antelope of either Africa or Asia.) He wrote: "That species of gazelle known as the antelope is very numerous upon the high plains. This beautiful animal, though reckoned a link between the deer and the goat, is certainly much nearest the latter. It is about the size and somewhat of the figure of a large goat. Its horns also resemble those of the latter . . . but they are more erect, and have a short prong projecting in front. . . . The antelope is most remarkable for its fleetness, not bounding like the deer, but skimming over the ground as though upon skates. The fastest horse will rarely overtake them. I once witnessed an effort to catch one that had a hind-leg broken, but it far out-stripped our fleetest 'buffalo horse.' It is, therefore, too swift to be hunted in the chase. I have seen dogs run after this animal, but they would soon stop and turn about, apparently much ashamed of being left so far behind."[8] It took the fastest dogs, such as the purebred greyhound, for hunters to be successful in the antelope chase.

Recounting the adventures he had on the Trail when he was eighteen years old, W. B. Napton included a chapter titled "My First Antelope" in his book *Over the Santa Fé Trail, 1857*. He recalled his attempts at flagging or toling the antelope: "After reaching the Cimarron [River], we began seeing herds of antelope in the distance. At first I tried 'flagging' them. I had been told that on approaching within two or three hundred yards of them, concealed from their view behind an intervening ridge, these animals were possessed of such an inordinate curiosity that they could be enticed to within gunshot of the hunter by tying a handkerchief on the end of a stick and elevating it in sight of the antelope, the hunter, of course, keeping concealed. I made several efforts at this plan of exciting their curiosity, and while some of them came toward me at first sight of the flag, their curiosity seemed counterbalanced by caution or incredulity, and in no instance could I get one to come near enough for a sure or safe shot. I then tried a rifle, with which I was also unsuccessful."[9]

Not one to give up easily, the persevering young hunter continued with his wagon train up the valley of the Cimarron, camping at each of the three springs called Lower, Middle, and Upper Springs as they followed the Trail. Napton recalled: "Captain Chiles had along with him two shot-guns, the smaller he had been using on buffalo, the other, an unusually large, double barrel, number 8 bore, very long in barrel and heavy, carrying easily twenty buck shot in each barrel. Armed with the big gun I would ride in the direction of the antelope, but at an angle indicating that I would pass them. Usually when I had gotten within three or four hundred yards of them they would quietly withdraw from view behind the ridge, whereat, I would turn the course of my horse and gallop as fast as I could, keeping the ridge between them and me until I had gotten within a short distance of the point of their disappearance. Then dismounting, I hastily followed them on foot. Often they would be found to not have moved out of the range of that big gun, and with it I killed many of them. That was the only plan of killing antelope by which I gained success."[10]

Before long, antelope became accustomed to hunters' efforts to attract them and grew increasingly wary of their intentions. One individual who noticed this change was Joseph Pratt Allyn, who wrote a letter at Raton Pass on the Mountain Route in November 1863: "We shot a stray rabbit and tried hard to kill an antelope, but those on the road are altogether too smart to be caught; they just keep out of rifle range and circle around you; they used to be more curious, and a red handkerchief or shirt would bring them within easy range."[11]

In June 1846, eighteen-year-old Susan Shelby Magoffin traveled the Trail in her own carriage close to the front of the caravan. She wrote in her diary, which was first published in 1926 as *Down the Santa Fe Trail and into Mexico*, while resting at midday at Bluff Creek Camp No. 8: "As we were quietly jog[g]ing along in front of the wagons, a beautiful little animal of some kind attracted our attentions. I supposed it a dog, or a wolf at first, my dearest [husband and veteran trader Samuel Magoffin] after many suppo-sitions settled on its being nothing more than a stone. To settle all doubt, we drew a spy-glass—and what was it? Nothing more or less than a timid though curious antelope. It did not run, but all curious as we were about it at first, to know what great objects we were coming toward it, it slowly advanced to meet us, but to its own destruction, poor creature!"[12]

Susan described how Samuel fired his rifle at the animal, and it jumped to the side of the road unharmed. However, as the curious creature continued its persistent gaze, he shot again and wounded the animal in its shoulder.

The antelope ran away, and Susan expressed her sorrow as she watched it flee from the scene: "It ran off over the hills, poor creature, no doubt to die. Since it was left with life and pain attached to it, I am sorry it was shot at."[13] The following month, she jotted in her diary: "We have seen several antelope. . . . It is a noble animal indeed; and there is certainly nothing that moves with more majestic pride, or with more apparent disdain [toward] inferior animals than he does. With his proud head raised aloft, nostrils expanded wide, he moves with all the lightness, ease, and grace imaginable."[14]

Marion Sloan Russell[15] also admired the graceful antelope, which she saw in great numbers during five trips over the Trail, beginning in 1852, when she was seven, with her mother, Eliza St. Claire Sloan, and brother Will Sloan. A decade later in 1862, at age seventeen, she made her fifth trip. Among the vivid memories of her Trail experiences included in her memoirs, published in *Land of Enchantment* (1954), was of the caravan heading out of camp on a summer morning: "There stretched out before us a new-coined day, a fresh minted world under a glorious turquoise sky. . . . The antelope stopped, stood still, and looked fearlessly at us. . . . I remember so clearly the beauty of the earth, and how, as we bore westward, the deer and the antelope bounded away from us. There were miles and miles of buffalo grass and blood-red sunsets and, once in a while, a little sod house on the lonely prairie—home of some hunter or trapper."[16]

J. W. Chatham of South Carolina wrote in his private journal in camp near Wagon Mound, New Mexico, on July 7, 1849: "I struck out with my gun for game, but did not succeed. Antelope is the principal [game], and it is amusing to see what inquisitive animals they are."[17] Earlier on the Trail near Pawnee Rock, he had briefly noted that an antelope fawn, about a month old, had been caught alive, and its tongue was coal black. Instead of hunting, Chatham collected mementos and items of interest found along his way over the Trail. He kept a list at the end of his journal. Among those mementos were "two female antelope horns from the Plains."

Not far from Wagon Mound was Fort Union, where Lieutenant William B. Lane was stationed in 1857. Impressed with the abundance of antelope in the area, he later recalled: "I could, from the front door of my quarters at the old post up against the bluffs at Fort Union, see at one time nearly any and every day several hundred antelope on the plain between the post and Turkey Mountain."[18] Today, pronghorn still live on the land where old Fort Union, now Fort Union National Monument, is located near Watrous, New Mexico. This National Monument and adjacent land are frequented by a few hundred pronghorn throughout the year, especially in late winter and early spring.

In 1863, newly married Ernestine Franke Huning traveled the Trail to Santa Fe and beyond to Albuquerque to live in the home of her husband, Franz Huning, a respected merchant who made forty trips over the Trail. An immigrant from Germany in 1849, Franz traveled to St. Louis and then to Fort Leavenworth, where he joined a wagon train as a bullwhacker. Ernestine refused to leave her beloved canaries behind and made the trip over the Trail in relative comfort, much as Susan Shelby Magoffin had on her trip, and even dined on goose breast and truffles, prepared by Franz's cook. She wrote a brief note in her diary in May 1863 about seeing many antelope, but her party had been unable to shoot any.[19] Some herders gave her and Franz a quarter of antelope for some salt, and she thought the meat tasted like venison.

While some disdained antelope meat, others on the Trail, like Ernestine Huning, thought it was quite palatable. On his way to the Colorado goldfields, Samuel D. Raymond recorded in his journal on May 23, 1859, that his caravan had completed eighteen miles along the Mountain Route and that he had seen the Spanish Peaks in the distance for the first time. These twin peaks were a guiding landmark for travelers once they reached southeastern Colorado. In good weather, they could be seen from about seventy miles away. He also noted: "Today we bought a quarter of antelope which was killed by two men encamped near us. We had some of it for supper, which was very good."[20] Another traveler, Hezekiah Brake, had emigrated from England to Minnesota and took the Trail to New Mexico in 1858 to work on a dairy farm west of Fort Union. He included in the "narrative" of his Trail experiences: "Mr. A[lexander] declared we were now within the range of antelope, and as we approached the Cimarron river we caught several glimpses of these shy and beautiful animals. As we neared our camping-ground, he was fortunate enough to bring down a fine young kid. When we had camped for breakfast, we took a sack of buffalo chips, carried forward for fuel, made our fire, and for the first time in my life I had the satisfaction of cooking and helping to eat fried antelope-chops."[21]

James Francis Riley's reminiscences of freighting on the Trail include an incident involving his brother and an antelope while they were on their way to Fort Union in the spring of 1863: "Brother Charley was one of the herders that night. He lay down close to an old steer and went to sleep. Just after it began to get light, he woke up and not more than a rod [16.5 feet] from him, stood an antelope looking at him as though it was wondering what he was. He had a little four-inch pistol in his pocket, slipped it out and shot and killed it. He said it made one jump and fell dead."[22] The shot woke up the rest of the guard and much excitement ensued over Charley shooting the antelope square between the eyes with his very small pistol. Riley added: "He [Charley] considered it a mere accident. But, we enjoyed the antelope all the same."[23] Although Riley, Brake, and others seemed to find antelope meat good eating, Josiah Gregg expressed a differing opinion: "The flesh of the antelope is, like that of the goat, rather coarse, and but little esteemed; consequently, no great efforts are made to take them."[24]

———

In recent times, only a few have written as eloquently about pronghorn as Jack Schaefer in his book *An American Bestiary* (1975). "My favorite [among the mammals who are confirmed vegetarians] is one who is a true native American," he wrote, "who to me is and always should be one of the prides of our continent, the pronghorn. . . . There is only one species of him. He is the one and only living representative of his genus, of his entire family. And he is completely and absolutely American."[25]

According to Schaefer, taxonomists were confounded with difficulties in classifying pronghorn, because this animal did not fit into any category. Few animals have caused so much debate about their place in the order of life. Naturalist Ernest Thompson Seton summarized the peculiarities that made their categorization difficult: "Like the giraffe, the American antelope [pronghorn] has two hooves [prominent toes] on each foot; like the goat, it has a gall bladder and a system of smell glands; like the deer, it has four teats and a coat of hair with an undercoat of wool; like the goat, it has hollow horns on a bony core; yet as in the deer, these horns are branched and are shed every year."[26] Seton did not mention that unlike the deer, pronghorn do not have dewclaws, which are the functionless digits on the lower legs or feet of some animals, such as the dewclaw on the inner side of a dog's legs or the small pair of dewclaws on the lower legs of deer. Also, like members of the large family of bovids, which includes cows, oxen, and goats, pronghorn are ruminants, or cud-chewers, with compartmentalized stomachs. They are ungulates (hoofed mammals), have high-crowned teeth for grazing, and they have horns, not antlers. Thus, pronghorn diverge from the bovid family and from all others to go their own way.

The horns of the pronghorn make this mammal truly unique. They are the only animal in the world with branched horns that are shed annually. Eventually accepting their singularity, taxonomists designated them *Antilocapra americana* and established their own family, Antilocapridae. Millions of years ago, the peculiarities of pronghorn had already been established. This species survived the extinctions of mammalian species over various epochs reaching back to the Miocene about twenty million years ago. Later in the early Pleistocene, pronghorn thrived along with mammoths, giant sloths, dire wolves, saber-toothed tigers, and other mammals roaming North America. Scientific studies have shown that more than ten or eleven thousand years ago, during the late Pleistocene, a large extinction or "a still mysterious die-up" occurred, possibly affecting about 70 percent of all species of large North American mammals.[27] All of the animals mentioned above became extinct, yet the pronghorn survived.

Among the pronghorn's greatest abilities are fleetness, commented on by many taking the Trail, and the most acute eyesight of any mammal. They are reported to be able to sustain sixty miles per hour for three to four minutes. Bursts of forty-five miles per hour are not unusual, and they are able to cruise at thirty miles per hour for up to fifteen miles.[28] Physiologically, pronghorn are perfect running machines, built for maximum efficiency. When running, they are almost level in action, which explains why Josiah Gregg thought they looked like they were "skimming over the ground as though upon skates."[29]

Some scientists believe their eyesight may be the best of any living thing, except for some predator birds and carrion-eating birds. Their large, wide-set eyes are about the same size as those of the elephant. Darkly pigmented eyes protect them from intense light from the sun and reflection from snow, and long, thick eyelashes provide shade. Set in protruding sockets, their eyes provide a wide arc of vision, with a power of sight reported to be "the equivalent of a human being with 20–20 vision using 8-power binoculars, giving them the ability to detect movement four miles away."[30]

Sharp eyesight is used in conjunction with their signaling system, or flashing of white rump patches. Pronghorn are able to communicate with one another over a distance of a mile or more. A specialized muscle system controls the coarse, loosely attached hair on their bodies, including rump patches, and allows the hair to stand almost erect or flatten layer upon layer. Besides providing a means of communication, it also provides excellent insulation in hot or cold weather. Scent glands are also used to communicate a variety of messages.

Pronghorn were living in North America when the modern buffalo, or bison, arrived from Asia, and they were here when the first human beings arrived in North America. As many as forty to fifty million pronghorn are thought to have once inhabited western North America from Canada south to the Mexican Plateau. Millions were on the plains when the first traders headed west from Missouri to Santa Fe in 1821. By 1915, however, an official government survey put the total estimated population for the United States and Canada at fifteen thousand.[31] Pronghorn populations suffered drastic declines during the nineteenth century and into the twentieth century for a number of reasons, but primarily because of loss of habitat, market hunting, and wanton slaughter. The population, however, began to show growth during the latter part of the twentieth century.

In 2013, the states with pronghorn had a combined estimated pronghorn population of more than 783,000. States with the largest estimated pronghorn populations were Wyoming (407,600), and Montana (110,225). Of the states through which the Santa Fe Trail crossed, Colorado and New Mexico had the largest estimated numbers of pronghorn in 2013, with 66,200 and 45,000 respectively, while Kansas had 2,750 and Oklahoma had 2,010. The estimated total for these four states is 115,960. Because Missouri is not in the range of pronghorn, the number for the state is zero.[32] These estimated pronghorn population counts were reported at the Twenty-sixth Biennial Pronghorn Workshop, which was held in May 2014 in Texas. The workshop, established in 1965, has been held every two years for interested parties from the United States and Canada. It is sanctioned by the Western Association of Fish and Wildlife Agencies and provides opportunities for the participants to meet and discuss issues, strategies, and research involving pronghorn, their conservation and management, and other important subjects. The population numbers for states and Canadian provinces are obtained through aerial and ground surveys.

It is revealing to compare the 2013 numbers with those reported for 2001 at the Twentieth Biennial Pronghorn Workshop (2002) for the four Trail states with pronghorn populations: Colorado, 54,070; New Mexico, 30,000; Kansas, 1,750; and Oklahoma, 1,200, with a total estimated population of 87,020.[33] Although population counts sometimes fluctuate during years the surveys have been conducted, the overall trend has been upward, even with drought affecting some parts of the participating states. Much of this success has been through the dedicated efforts of wildlife departments, conservation and research organizations, and a number of private citizens, including ranchers and hunters. All of these people's cooperation and heroic efforts in habitat enhancement, in particular, have been effective. A moratorium on hunting pronghorn lasting until the 1940s and the passage of laws to raise money for wildlife restoration and other conservation efforts were also instrumental. Valerius Geist, who writes and speaks about the pronghorn, has stated: "The pronghorn is now abundant because we made it so. . . . Today, the pronghorn stands as a symbol to much that is good and decent in North America."[34]

Over the ages, pronghorn adapted to life on vast, open spaces, learning to rely on their speed, agility, and sharp eyesight to outrun any predator. Unlike deer and other wild animals, they do not seek cover, flee, or run blindly from potential danger, but respond to predators selectively. It is their natural response to check out whatever may be a threat. After scoping out what has attracted their attention, they will either run or not run. Frequently, they will stay where they are and totally ignore whatever it was that had caused such close scrutiny. Thus, what observers have called "curiosity" is the pronghorn's natural behavior in responding to possible threats. This response had worked well for them. With the arrival of predators with guns, however, pronghorn could no longer rely on their instincts and swiftness to outrun such a dangerous threat as a speeding bullet.

As you hike, ride, bike, or drive along the Santa Fe National Historic Trail or open spaces of the American West and are fortunate to catch sight of pronghorn, recall their ancient heritage, long history on the prairies and plains, and near extinction. Remember that these colorful, graceful, fleet animals are survivors—and true Americans.

Chapter Three

Prairie Dogs

Santa Fe Trail travelers were fascinated by prairie dogs. These social animals belong to the family of rodents, Sciuridae, which includes squirrels, chipmunks, and marmots. Living in close-knit family groups clustered together in large colonies, they aroused the curiosity of the travelers following the Trail over the grasslands and along the Arkansas River. They frequently recorded observations of prairie dogs in their journals and diaries in greater detail than for any other animal, with the exception of the buffalo. Travelers wrote about prairie dogs' physical characteristics, behavior, burrows, and "towns," and about the name of this animal, which seemed a misnomer to many.

Men wrote about the difficulty in shooting these quick-moving animals and that they must be shot dead, because if only wounded, the prairie dogs would drop into their burrows, leaving no opportunity to retrieve them. Some observers were convinced that prairie dogs, rattlesnakes, and burrowing owls resided together in some sort of mutual arrangement in the "villages," while others contended that snakes and owls occupied the burrows only after they were abandoned by prairie dogs.

Prairie dogs did not pose the threat that many travelers felt from buffalo, wolves, and rattlesnakes, or cause the agony inflicted by such insect pests as mosquitoes, buffalo gnats, fleas, flies, lice, mites, and ticks. The prairie dogs' amazing behavior and humorous antics provided a much-welcome diversion and some comic relief for the travel-weary sojourners. The travelers seemed to have a special feeling toward these sociable animals, frequently describing them in a variety of human terms. They wrote about the dogs' "houses" and their "towns" or "villages" with "streets." They also wrote about the dogs' neighborliness, their caring for each other, their apparent conferences described as a "grand council" (also a "confab," "consultation," or

"gossiping"), and their watchfulness and warnings to others of approaching danger. The tight-knit families of prairie dogs may have caused the travelers to think of their own families far away and their longing to be with loved ones, of the warmth and security of hearth and home, and of their own need to be ever vigilant of the many types of danger about them on their long journeys over the Trail. Their comments showed a softness of heart for these grassland squirrels.

In a series of articles about his Trail experiences in 1839, writer-actor Matthew (Matt) C. Field penned one titled "Dog Towns." He stated that these towns were "one of the most striking peculiarities which rivet attention in the buffalo regions. . . . These are spots of short grass, growing exceedingly thick and fine."[1] Matt also explained how the animals dug out their houses, or burrows, and made brown earthen mounds around the entrances, "spangling the darker green of the prairie in a manner that would doubtless be exceedingly picturesque could it be viewed from a balloon."[2]

Traveling to Arizona Territory in the 1860s to take a position as an associate justice, Joseph Pratt Allyn wrote a letter in which he included the following observations about prairie dogs: "Just imagine miles of level, barren prairie, covered with rather large ant hills without any regularity of arrangement, and you have the external appearance of a prairie-dog town. If you have a good eye or a good glass [spyglass] you may see some of the dogs a long way off. They are about the color of the ground, and the size of a rabbit, and somewhat the shape of a squirrel. Why they are called dogs is more than I can imagine. The owls are more casually seen, and once in a while a rattlesnake drags himself out and shakes his rattles. We thrust sabers and fired pistols into the holes and never startled anything out. Whether the owl, dog and rattlesnake live in the same hole is, I suspect, open to very grave doubt. They certainly live in the same town, but as these are miles long it doesn't necessarily imply joint tenancy of the holes."[3]

Josiah Gregg saw countless prairie dogs on his four trips over the Santa Fe Trail from 1831 to 1840. He wrote several long paragraphs about these "curious" animals in *Commerce of the Prairies* (1844). His first paragraph shows that he had observed them closely and was also captivated by them: "But of all the prairie animals, by far the most curious, and by no means the least celebrated, is the little prairie dog. This singular quadruped is but little larger than a common squirrel, its body being nearly a foot long, with a tail of three or four inches. The color ranges from brown to a dirty yellow. The flesh, though often eaten by travelers, is not esteemed savory. It is denominated the 'barking squirrel,' the 'prairie ground-squirrel,' etc., by early explorers, with much more apparent propriety than the present established name. Its yelp, which resembles that of the little toy dog, seems its only canine attribute. It rather appears to occupy a middle ground betwixt the rabbit and squirrel—like the former in feeding and burrowing—like the latter in frisking, flirting, sitting erect, and somewhat so in its barking."[4]

Gregg's lively description of the little prairie dogs' behavior provides a good example of observers' inclination to attribute human characteristics to these animals: "Approaching a 'village,' the little dogs may be observed frisking about the 'streets'—passing from dwelling to dwelling apparently on visits, sometimes a few clustered together in council—here feeding upon the tender herbage—there cleansing their 'houses,' or brushing the little hillock about the door—yet all quiet. Upon seeing a stranger, however, each streaks to its home, but is apt to stop at the entrance, and spread the general alarm by a succession of shrill yelps, usually sitting erect. Yet at the report of a gun or the too near approach of the visitor, they dart down and are seen no more till the cause of alarm seems to have disappeared."[5] ✓

Lieutenant William Fairholme of the British army wrote about prairie dogs along the Trail in his "Journal of an Expedition to the Grand Prairies of the Missouri, 1840." His experience with these animals and a rattlesnake occurred near the Pawnee Fork of the Arkansas River while pursuing buffalo with two companions (Jean Baptiste Tabaud and a servant known only as Pierre): "We rode on together for some distance when I perceived, on a bare sandy spot, some little animals like rabbits running to & fro, as if in a great bustle, and others sitting up on their hind legs. I got off my horse, and followed by Pierre, stole quietly up to get a shot at them, but just as I was taking aim at three of them, altogether a beautiful shot, I was startled by a loud cry from Pierre.

"The prairie dogs, as they are called, disappeared instantly, their heels twinkling for a moment in the air as they dived into their holes, and I looked round to Pierre, who screamed out, 'Le serpent! Gardez-vous du serpent!' And there, sure enough, close at my feet was a large rattlesnake advancing toward me with his head up in the air, and his eyes sparkling like diamonds. I made off immediately, and when a few paces off, waited to see what he would do. He came on in the most graceful undulations till he arrived at the mouth of one of the prairie dog burrows, which he seemed inclined to enter. But as he had spoiled my sport, I took advantage of his pause to take a shot at him. My ball cut his head clean off, and his body remained twisting & writhing for a long time afterward.

"I sat down with Pierre to wait till the little creatures should show themselves again, and presently one popped his head out of a hole, then another and they seemed to hold a consultation together for a short time, when they both came out and sat up on their hind legs in the most comical way, opposite each other, till two or three others joined them. It was really very amusing to watch the bustling manner as they all ran about from hole to hole, as if paying visits of congratulation on the danger being past, for they could not see us, we being concealed behind a thick bush of grass. I put an end to their gossiping after I watched them for a little time by firing a ball into the middle of them (having no shot with me). I was almost sorry to see that I had severely wounded one which, just like a rabbit, managed to scramble into a hole close to him, and after a fruitless attempt to get him, we remounted our horses and pursued our route."[6]

Fairholme thought prairie dogs might be a species of marmot. He mentioned the little owl, which lives in abandoned burrows, and also rattlesnakes that he frequently saw entering and leaving the burrows. Many travelers thought the three lived together, but Fairholme and others were certain that was not true. He commented: "If it is the case, it is a remarkable fact that a bird, a serpent & a quadruped should inhabit the same den, more particularly as I should imagine there can be no doubt that the rattlesnake feeds on the prairie dog."[7] Fairholme was correct. Prairie dogs are in no way friends with rattlesnakes. Other animals in addition to snakes have depended on them and their young for food. Their old, abandoned burrows also provide welcome shelter and habitat for other creatures, such as the burrowing owl (*Speotyto cunicularia*), the small, long-legged diurnal owl of the plains.

Lewis Garrard was charmed by the little "dogs" as he rode through their "villages." He recalled those experiences in *Wah-to-yah and the Taos Trail*, including in his account what he had "ascertained through inquiry and observation." The day after reaching "the grand Arkansas," Garrard described the river as "quite broad, with two feet of water, sandy bottom, and high sand buttes on either bank, as bare & cheerless as any misanthrope could wish," and recounted his observations after riding through two "towns": "These villages are frequent; often we came across them several miles from water, but whether they abstain from it totally is a question not solvable by any mountain men with us. Very little or no dew falls in this region, so these strange animals do not depend upon this source to quench their thirst. . . . The grass in this region is short, early, and highly nutritious. It has a withered, brown appearance even early in the spring, and is designated as 'buffalo grass.'"[8] In fact, the grasses, herbaceous plants, insects, and grubs eaten by them usually provide enough water to meet their needs.

Like many on the Trail, Garrard found the prairie dogs entertaining: "It is quite amusing to watch their movements on top of the cones [the entrance mounds can be as tall as four feet high]; on our approach, they barked, their short tails nervously fluttering, and receiving a new impetus from the short, quick, and sharp tiny yelp which they constantly uttered; when they thought themselves in danger, with an incredibly quick motion, they threw themselves back in the holes and immediately reappeared with an impertinent, daring bark, as if to say, 'you can't get *me.*' Others 'crawfished,' hiding by their singular way of crouching the back, until nothing but their heads and tails could be seen—these latter shaking tremulously. Succeeding a silence of a few minutes after scaring in the 'dogs,' we could see by lying flat on the ground so as to get the tops of the cones between the sky and our eyes, with the closest scrutiny, the head, here and there, of a dog almost imperceptibly moving, and with a cautious reconnoiter to see if the coast be clear, he would show himself and with a knowing yelp apprise his neighbors of his investigations."[9]

Some travelers made note of their first sighting of prairie dogs; many had already read or heard about them, particularly Josiah Gregg's description of them in *Commerce of the Prairies.* One of those readers was Philip Gooch Ferguson, company clerk of the First Regiment of the Missouri Mounted Volunteers in the Army of the West. On July 25, 1847, Ferguson wrote in camp near the "swift, muddy stream" called Pawnee Fork: "Today I saw for the first time a village of prairie dogs, an animal whose habits have excited the wonder of man and the admiration of the lovers of natural history. The prairie dog is but little larger than a full-grown fox squirrel, of a reddish brown color, like a ground hog—short tail, small ears, teeth and head like a squirrel's or rat's, and very compactly shaped. Their villages are frequently several miles in area, their houses being holes running slanting in the ground, the dirt around the orifice sometimes rising several feet above the surface like a chimney. Around the outskirts of the village, sentinels are posted, who on the approach of danger, give the alarm by their peculiar bark, from which doubtless they derive their name of 'dog,' although it is more like the bark of a squirrel, there being but little resemblance between them and the dog."[10]

Ferguson noted in his diary the following night after marching twenty-two miles and making camp on the Arkansas: "The men had been shooting at prairie dogs all day, and not killing one, they concluded that the dogs 'dodged the flash,' but I do not think this can be so. But it is said they are very hard to kill and unless struck instantly dead, will tumble over and get into their holes."[11]

Susan Shelby Magoffin recorded in her diary seeing her first prairie dog town on July 2, 1846: "Prairie scenes are rather changing today. We are coming more into the buffalo regions. The grass is much shorter and finer. The plains are cut up by winding paths and every thing promises a buffalo dinner on the 4th [Independence Day]. We left our last night's camp [Cow Creek] quite early this morning. About 9 o'clock we came upon 'Dog City.' This curiosity is well worth seeing. The Prairie dog, not much larger than a well grown rat, burrows in the ground. They generally make a regular town of it, each one making his house by digging a hole, and heaping the dirt around the mouth of it. Two are generally built together in a neighborly way. They of course visit as regularly as man. When we got into this one, on both sides of the road occupying at least a circle of some hundred yards, the little fellows like people ran to their doors to see the passing crowd. They could be seen all around with their heads poked out, and expressing their opinions I supposed from the loud barking I heard."[12]

Frank Edwards, who also served in the Army of the West, marched the length of the Santa Fe Trail in 1846 and saw incredible numbers of the little prairie dog. He described the "towns" and their inhabitants: "We passed by and over several prairie dog towns. One of these was very extensive, being three or four miles in circumference, and the ground shook under us as we crossed it, with a hollow sound, as if we were passing over a bridge. Although the name of dog is applied to these little animals, they bear no possible resemblance to our dogs, even their cry is most like a bird's chirp. They are much smaller than generally represented, being a trifle less in size than the common rabbit, and far superior to the latter in flavor."[13]

One of the most detailed descriptions of prairie dogs by a Trail traveler was provided by W. W. H. [William Watts Hart] Davis in his book *El Gringo; New Mexico and Her People* (1857), which is an account of his 1853 trip by stagecoach to New Mexico Territory, where he served as U.S. attorney. He recalled the time his party observed them in the area of the Oklahoma Panhandle on the Cimarron Route: "We were now traveling through the region inhabited by the 'prairie dog,' and the whole country seemed one continued village. They are a curious and interesting little animal, and deserve a passing notice. For miles the plains are dotted with the piles of dirt before their holes, which resemble large ant-hills. They dig a deep hole in the ground, four or six inches in diameter, and carry up the dirt and place it in a heap at the mouth in the shape of a cone, and about a foot high. Their holes are unequal distances apart, and are arranged without order. . . .

"Those that first discerned your approach seem to have been sentinels, stationed to sound the alarm to the main body. Now the town is aroused, and every able-bodied citizen comes out of his hole to be prepared for any emergency that may arise. As you approach nearer[,] their activity increases, and frequent communication is held between different quarters of the town. Now you notice three or four in close conclave, as if holding council upon the affairs of the nation, at the end of which they separate, each one returning to his own home. Now you observe a single dog run across to his neighbor, hold a moment's confab with him, and then skip back again. In another part of the village you will see them assembled in grand council, in considerable numbers, apparently holding a solemn debate upon the state of public affairs. They are formed in a circle, each one sitting erect upon his hind legs, and in the middle is seated a grave old patriarch, who has the required wisdom to preside over and direct their deliberations. Apparently some important question has been discussed and decided, for, when they adjourn, messengers are seen hastening to all parts of the town to announce the results. Thus the little rascals keep up their operations until you draw very near, when every fellow disappears in his hole, and you see nothing more of them while you remain in the village. . . . They are seldom caught, and will not even leave their holes when water is poured in upon them."[14]

Others attempted, without success, to flush prairie dogs out of their tunnels with water. In July 1851, Catherine (Katie) Carey Bowen, accompanying her husband, Captain Isaac Bowen, from Fort Leavenworth to Fort Union via the Santa Fe Trail, wrote at their camp near Cow Creek: "For several days we have been passing through 'dog towns.' They cover acres and acres, little holes a few feet apart and deeper than anyone knows. We tried this morning to drown some out and poured many buckets full of water into their holes without any success. While running down the water sounded twenty or thirty feet below the surface."[15]

Decades earlier, in 1806, Captain Zebulon Montgomery Pike and his exploring party attempted to flood prairie dogs out of their burrows. He wrote in his journal that he thought the prairie dogs' burrows descended in a spiral form, which he surmised was the reason he could not ascertain their depth. Pike wrote: "I once had 140 kettles of water poured into one of them in order to drive out the occupant, but without effect."[16]

Two years before Pike's experiment, the Corps of Discovery, including Captain Meriwether Lewis and Co-captain William Clark, tried to flush some prairie dogs out of their burrows. On September 7, 1804, the first day they encountered prairie dogs, the men, except for a few guards, halted their expedition "for the better part of a hot day" to try to dig some prairie dogs out of their burrow and capture one alive to send as a gift to President Thomas Jefferson. When that failed, they attempted to flood one of the animals from its hole. After great effort, they were able to catch one prairie dog. The following March, Lewis and Clark sent President Jefferson a large collection of plant and animal specimens gathered along their way over the northern plains. The only live animals included in the cargo were the solitary prairie dog, four magpies, and one "living hen of the prairies."[17] The hen of the prairies was some type of grouse—possibly sage grouse, sharp-tailed grouse, or prairie chicken.

The grueling journey left from Fort Mandan (located in present-day North Dakota), headed down the Missouri River to the Mississippi, south to the port of New Orleans, by ship around Florida and up the Atlantic coast, and finally overland to Washington, D.C., arriving on August 12, 1805. The prairie dog survived the trip, but of the five birds only one magpie was alive. According to accounts, President Jefferson was very pleased with the prairie dog and sent it to Philadelphia to be kept on display at the natural science museum inside Independence Hall. In addition to being popular at the White House, the prairie dog aroused the curiosity and interest of the public. It is possible that live prairie dogs made trips in cages or crates over the Santa Fe Trail, but no record of such an occurrence was found.

———

Captain Pike was among the first to describe the prairie dog in an official published document. His description of the black-tailed prairie dog is considered his main contribution to the world of natural history.[18] This account was based on his observations of the prairie dogs and their towns scattered over the vast territory of the middle plains through which the Arkansas River coursed. Those observations were made in 1806 during Pike's expedition to the Southwest, one with a long list of objectives, including ascending the Arkansas to its source, and an important factor in the opening of the Santa Fe Trail.

Thinking these animals were more like squirrels than any other animal he had seen, Pike preferred the name "prairie squirrel," which he considered more appropriate than "prairie dog." Others agreed with Pike's view, but the common name stuck. W. Eugene Hollon remarked in his book *The Lost Pathfinder, Zebulon Montgomery Pike* (1949): "The contrast between a squirrel that climbed trees in the East and one that burrowed in the ground in the West was too much for most travelers who crossed the plains after Pike. Consequently, the prairie squirrel was made a dog!"[19] Hollon added that prairie dogs were called *petit chiens* (little dogs) by the French. Spaniards called them *tuza*, a name derived from Nahuatl, the language of the Aztecs.

Pike wrote a lengthy account about the prairie dog's habits in his journal on October 24, 1806: "The Wishtonwish of the Indians, prairie dog to some travelers, or prairie squirrel as I should be inclined to denominate them, reside on the prairies of Louisiana [the Louisiana Territory, purchased from the French in 1803] in towns and villages, having an evident police established in their communities. . . . Their residence, being underground, is burrowed out and the earth [mound] answers the double purpose of keeping out the water and affording an elevated place in wet seasons to repose on, and to give them a further and more distinct view of the country. . . . As you approach their towns, you are saluted on all sides by the cry of Wishtonwish, from which they derive their name with the Indians, uttered in a shrill and piercing manner. . . . It is extremely dangerous to pass through their towns, as they abound with rattle snakes . . . and strange as it may appear, I have seen the Wishtonwish, the rattle snake and other animals take refuge in the same home. I do not pretend to assert that it was their common place of resort, but I have witnessed the facts more than in one instance."[20]

Physical and other characteristics of the black-tailed prairie dog were described by Pike: "They are of dark brown color, except their bellies, which are white. Their tails are not so long as those of our grey squirrels, but are shaped precisely like theirs; their teeth, head, nails, and body, are the perfect squirrel, except that they are generally fatter than that animal. Their villages sometimes extend over two and three miles square, in which there must be innumerable hosts of them, as there is generally a burrow every ten steps in which there are two or more, and you see new ones [burrows] partly excavated on all borders of the town. We killed great numbers of them with our rifles and found them excellent meat, after they were exposed a night or two to the frost, by which means the rankness acquired by their subterraneous dwelling is corrected."[21]

Although there were differing opinions regarding the savoriness of the meat, prairie dogs would often become a substitute when other meat was scarce or not available, ending up in someone's soup pot or dressed for roasting over an open fire. J. W. Chatham of South Carolina briefly commented in his Trail journal in 1849: "Having traveled about twenty miles, camped tonight on the bank of the Arkansas, a muddy and shallow stream. One of the hunters brought in a prairie dog—a sorry looking puppy—for soup."[22]

Such was the fate of a prairie dog when Hezekiah Brake left Minnesota and traveled the Trail in 1858 to take a job as manager of a dairy farm west of Fort Union. Brake recounted in his diary: "We were passing through a prairie-dog village. At the door of his habitation, a fierce young dog set up a yelp of remonstrance at our interrupting their councils, and Mr. A[lexander] silenced him with a bullet. Throwing the dog into the wagon, we went on to our limit of fifteen miles, and stopped for breakfast. I had cooked 'possum, 'coon, even terrapin, in my time, and was not to be deterred by jeers from preparing fresh meat simply because there was no material at hand to cook but a prairie dog, and no fuel for a fire but buffalo chips. So I made my fire, put a vessel of water on to boil, and dressed the dog. A savory stew was soon prepared which threw fried bacon into the shade. All of us pronounced prairie dog superior to squirrel or rabbit, and declared that after this we would often have fresh meat."[23]

Brake's admiration of prairie dog meat, however, was short-lived, and his opinion of bacon was restored when he prepared another "dog" the following day. The party rose early the following morning, cleaned and greased the wagon axles, and started out on their day's journey. That evening in camp near Pawnee Fork, Brake wrote: "Without breakfasting, made twenty miles.

By this time we were again hungry for fresh meat. Mr. A. shot a fat young prairie dog as before, and I skinned the animal and prepared him for the pot. Being very lean myself, I have always been a great admirer of fat, and I testified to this admiration now by putting a piece of the stuff into my mouth. I had no more than masticated and swallowed that piece of fat until I was sicker than words can express. In disgust, I threw the whole dog away, and I have never since particularly cared for prairie dog meat. As to the fat, that mouthful has lasted me through all of the years that have since elapsed. It took a strong cup of coffee to cure the dog fit from which we all seemed to be suffering, and bacon and eggs tasted like ambrosia."[24]

Prairie dog meat became a fast food for those who did not have time to prepare buffalo or other large animals. James Josiah Webb, a well-known merchant on the Santa Fe and Chihuahua Trails in the 1840s, recorded a memorable episode from his early experiences as "an unmitigated greenhorn." His group was headed west toward the mountains after leaving Bent's Fort. While in camp early one afternoon, the group saw three men approaching at a brisk gallop. Webb recalled, "In a few minutes they had a fire kindled, and the coffeepot over the fire. They were soon recognized as old mountain men and acquaintances of several of [our] party—Kit Carson, Lucien Maxwell [holder of the huge Maxwell Land Grant], and Timothy Goodale [a trapper and guide]. As soon as they got dinner cooking (coffee boiling, a prairie dog dressed and opened out on a stick before the fire), Carson and Maxwell came to our camp. This was my first interview with these celebrities. It was very short, and I can remember nothing of the interview except that they left Pueblo [Colorado] that morning and expected to reach Taos that night. They soon left, ate their dinner, saddled their horses, caught their led horses, and were off. . . ."[25]

Prairie dog meat was also eaten on festive occasions. In December 1858, Edward F. Beale and his party were headed toward winter quarters at Hatch's Ranch in New Mexico. At that time, Beale was an important figure in the American West, but he is primarily remembered for his efforts to introduce camels into the Southwest and for charting a wagon road almost due west from Fort Smith, Arkansas, following the course of the Canadian River. After entering New Mexico, the party stopped at a place called Laguna Colorado to spend Christmas Day and enjoyed a feast that included wild turkey, deer, antelope, raccoon, grouse (quite possibly prairie chicken in this location), and prairie dog.[26]

———

Each prairie dog has its place in the family, and each family plays a role in the life of the town, or colony. Similar to a human city, a colony is organized. It may be divided up into smaller sections called wards. Within each ward, prairie dogs live in small family groups called coteries. A coterie typically consists of one adult male, several (usually three or four) close-kin adult females, year-old offspring, and new young of the season. Each coterie shares a burrow system that covers about 0.6 to 1.0 acre. Their burrows may be as deep as sixteen feet and have tunnels from twelve to one hundred feet long.[27] Along tunnel walls, holes for rooms, or chambers, are scooped out and used for a variety of purposes from nesting to burying trash. The social lives of these animals caused many travelers to write about the ways they seemed to be communicating with each other, such as in "confabs" and "councils." Recent research relating to their alarm calls has found that they discriminate among predators, even people of different sizes. They have been called "talented linguists."[28]

Of the five species of prairie dogs inhabiting western North America—black-tailed, white-tailed, Gunnison's, Utah, and Mexican—two have ranges that include the routes of the Santa Fe Trail. The black-tailed prairie dog (*Cynomys ludovicianus*), the most common and most social of the five species, lives on short and mixed grasslands, once stretching the length of the plains from southern Canada to northern Mexico. The tall-grass prairie, where predators can hide, limits its range to the east. Black-tailed prairie dogs ranged along the routes of the Santa Fe Trail from central Kansas through southeastern Colorado, across the Oklahoma Panhandle, and into north-eastern New Mexico. This is the species that Santa Fe Trail travelers saw and wrote about during their journeys across the prairies and along the Arkansas River.

As the plains turn to foothills and mountains in Colorado and New Mexico, the range of Gunnison's prairie dogs (*Cynomys gunnisoni*) begins and extends westward through the Four Corners area. It also extends south from northern New Mexico through Santa Fe to the central part of the state. Biologists estimate that at the end of the 1800s as many as five billion prairie dogs occupied millions of acres. Today, prairie dogs inhabit only about 2 percent of their former range, or historic habitat, from Canada to Mexico.[29]

As settlers moved onto the plains, turning grasslands into croplands, and as cattle filled the void where buffalo once roamed, the softness of heart toward the little prairie dogs turned to a hardness of heart. No longer considered a wonder of natural history, they became unwanted, vilified vermin to be exterminated at any cost. Decades of poisoning, habitat destruction, recreational shooting, and eradication efforts by government agencies drastically reduced the population and range of prairie dogs. Since the 1940s, sylvatic plague, a flea-borne disease that arrived in North America in 1900, also became a factor in their decline, because they have little or no immunity to it.

For thousands of years, prairie dogs shared their home, the prairie grasslands, with millions of buffalo, pronghorn, and numerous other grassland creatures. These animals had long-established mutual relationships. Grazing buffalo, for example, left large areas of close-cropped grasses, which prairie dogs prefer because they can more easily see their numerous predators. In turn, prairie dogs benefited buffalo and other grazing animals by digging up and loosening the ground, bringing up fertile subsoil and minerals, and aerating the earth. Rainwater could soak into the ground more easily. They also fertilized the soil. All of this activity encouraged the growth of stronger, healthier, and more nutritious grasses and other plant life. As prey, prairie dogs benefited many other animals, and their abandoned burrows provided habitat and shelter.

In the latter part of the twentieth century, scientific research began to show the contributions of prairie dogs to a well-balanced prairie ecosystem. It was also found that the long-term programs to eradicate them through the use of poisons were taking a terrible toll on dozens of other species linked to the prairie dog. The black-footed ferret, North America's most endangered mammal, is a prime example of the more than one hundred species, including mammals, reptiles, amphibians, birds, and insects, that either depend on the prairie dog for food, shelter, and habitat or benefit from its activities. The black-footed ferrets' primary source of food has long been the black-tailed prairie dog. This ferret, once declared extinct, was rediscovered in 1981. Captive breeding has shown promise in its recovery.

In addition to the black-footed ferret, other threatened or endangered species that are most tied to prairie dogs for survival include swift fox, burrowing owl, golden eagle, mountain plover, ferruginous hawk, horned lark, grasshopper mouse, and deer mouse.[30] Of the species of mammals benefited by the black-tailed prairie dog, many are the mainstays of the Great Plains and icons of the Old West—the animals that once awed and delighted travelers on the Santa Fe Trail.

Public and private entities, alliances, and concerned citizens have been striving to find "a middle ground" and come to grips with issues surrounding the prairie dog. Their common vision is to work toward restoration of the prairie dog ecosystem in focus areas of several states of the Great Plains. For example, New Mexico established a state management plan aimed at increasing the acreage on which black-tailed prairie dogs live. This plan has involved a mix of public and private participants and emphasized the voluntary cooperation of landowners. Several ranches in the state began receiving prairie dogs requiring relocation after being removed from their former habitats, in particular from expanding urban areas. Another example is the purchase of land in South Dakota by The Nature Conservancy to protect vital habitat for both black-tailed prairie dogs and black-footed ferrets. Research is continuing on the management of prairie dogs. Such concerted efforts are very important in giving the prairie dog, the recognized keystone species of the short-grass prairie ecosystem, new and more permanent homes, and giving this prairie ecosystem a chance at regeneration in some protected areas of our historic American West.

Chapter Four

Wolves

Although the buffalo was the largest of the animals seen by the travelers on the Santa Fe Trail, it was not considered to be the ruler of the other animals. The "scepter of authority," according to Josiah Gregg, clearly belonged to the gray wolf.[1] No one on the Trail disputed Gregg's statement, because they saw with their own eyes the authority wielded by wolves. He described this wild member of the dog family Canidae as "ferocious" and told how these powerful animals cooperated in hunting and killing their prey. Formidable predators, they roamed great distances, hunting buffalo and other large mammals.

The gray wolf (*Canis lupus*) was emblematic of the open prairies and plains. Like the buffalo, pronghorn, prairie dogs, and other animals, wolves were found in incredible numbers and especially on the buffalo ranges. Also like the buffalo, they were almost hunted to extinction. This species was once common throughout almost all of North America. As the number of wolves increased along the westward Trail, travelers knew they were approaching buffalo territory and would soon catch sight of the great multitude of herds.

Greenhorns who joined the trading caravans on the Santa Fe Trail may have started with little knowledge of the differences between wolves and coyotes, but they soon could distinguish the two species easily. Living and hunting in family groups, or packs, wolves have a social order called a dominance hierarchy. The wolf pack is a closely knit unit headed by an alpha, or lead, pair. One or more subordinate wolves, usually the offspring of the alpha pair from previous years and the new brood, or youngest offspring, of the current year constitute the rest of the pack. The subordinate members participate in the hunt and in raising the new brood, but they rarely breed. The size of a pack is related to prey size. Adult males weigh from 75 to 110 pounds, measure five feet to six and one-half feet long (tail included), and stand two and one-half feet tall at the shoulder. Adult females are smaller than males. Features distinguishing wolves from coyotes are larger body size, larger nose pad, shorter ears, and a longer tail often tipped in black and usually held horizontally (not down) when running.[2]

Lured by the account of pathfinder John Charles Frémont's tour to the Rockies in 1842 and 1843, seventeen-year-old Lewis Garrard started out as a greenhorn on the Trail in 1846. With youthful exuberance, he quickly took to the life of the frontiersman and became a full-fledged member of a buffalo-hunting party—that is, once he was able to figure out how to catch his horse. He described a scene of buffalo chased by wolves: "Our attention was drawn to a band of buffalo running across our path, a half-mile to the rear, and two hundred or more large wolves, who, with outstretched necks and uplifted sharp heads, were in sure, noiseless, though swift pursuit. It was a magnificent sight to watch them dashing along—the poor buffalo straining their utmost to elude the sharp fangs of their persecutors—the wolves gaining at every stride. On they went, now out of sight, now in the river, where the buffalo had the advantage; a cool swim invigorated the pursuers, who, loping with dripping hair, howled, as they pressed on to victory."[3]

Garrard also recalled the voracity of the wolves after the hunters butchered a buffalo where it had fallen: "Loading our animals with choice pieces of the tender cow, we left for the Trail, much to the apparent satisfaction of some wolves, loping and howling or sitting on their haunches, seemingly resolved to bide their time. Looking back, after we left a short distance, we saw them fighting, with their tails whisking about quite lively in the struggle for 'spoils.'"[4]

Many travelers on the Trail wrote about wolves following the caravans. They usually loped along behind the wagons at a safe distance, having quickly learned to stay out of rifle range. They would also hang around close to the camps and the travelers' wagons. Lieutenant James William Abert, a West Point graduate assigned the task of preparing a scientific report about New Mexico for the Corps of Topographical Engineers, accompanied the Army of the West over the Trail in 1846. He wrote on September 18: "Great numbers of wolves, the large grey wolf, were prowling around our camp."[5]

In 1850, James Cleminson, traveling with his wife, Lydia Ann, and their six children in a wagon train of emigrants headed over the Trail from

Independence, Missouri, wrote in his journal at Lower Cimarron Spring on August 31: "There are on the plains a great number of wolves. They are around us every day more or less, and sometimes come quite near to our wagons."[6] The children in the caravans were taught to stay close to the wagons and never wander off. Marion Russell mentioned that she and her brother were always reminded by their mother to be cautious and stay with the wagons, unless an adult was with them. Children were always counted before the caravan headed out each morning from camp.

Dr. Michael Steck wrote a letter on December 11, 1852, after reaching Santa Fe. He had left Independence, Missouri, on October 10 to assume his new post in New Mexico as agent for the Mescalero Apaches. His party of twenty-three people traveled in three mule-drawn carriages with three wagons for their baggage and provisions and several animals for riding. He described their progress toward Fort Atkinson, located on the Arkansas River west of present Dodge City, Kansas: "We continued on . . . without anything to interrupt us except an occasional squall of snow. We killed several Buffalo and occasionally a Wolf when we thought he showed too much impudence. They [the wolves] are generally shy, but we see immense numbers of them. A common thing to see [is] fifty at a sight and in the region of the Buffalo. In the daytime, never out of sight of them, see hundreds in a day. They live upon the Buffalo, the [calves] and old ones. They select [one to run down], surround it and keep snapping it in the hind legs until they hamstring it. This accomplished, they get him down and frequently in an hour devour the largest Buffalo."[7]

In 1866, many wolves were still chasing the buffalo, a scene witnessed by William N. Byers, co-publisher of the newspaper *Rocky Mountain News*. He wrote at Fort Aubry, a temporary post used to guard the Trail in western Kansas from 1865 to 1866: "Wolves follow and hang about the buffalo herds in incredible numbers."[8] Byers had arrived at the fort in an eastbound stagecoach from Bent's Fort in January, after traveling in "excessively cold" and snowy weather, which had forced wagon trains on the Trail to lay up until the weather improved.

A number of people taking the Trail mentioned the presence of white wolves on the plains, even packs of white wolves. Although gray was the predominant color, wolves range from shades of gray, brown, and tan, and may also be solid black or pure white. One of the people mentioning white wolves was Bostonian Albert Pike, who traveled in a caravan led by Charles Bent in 1831, a decade after the opening of the Trail to international commerce. Pike also wrote about the voraciousness of wolves: "Our oxen, from hunger and drought, began to fail, and we were, every day or two, obliged to leave one behind us. The hungry jaws of the white wolves soon caused them to disappear from the face of the earth, and by thus affording these voracious animals food, we had a continual train of lean, lank and gaunt followers, resembling Hunger-demons, following us stealthily by day, and howling around us by night. . . . Our oxen were daily decreasing in number, and our train of wolves enlarging. I can give the reader some idea of their number and voracity by informing him that one night, just at sunset, we killed six buffaloes, and having time to butcher and take to camp only three, we left the other three on the ground, skinned and in part cut up. The next morning there was not a hide, a bone, or a bit of meat, within fifty yards of the place."[9]

James Ross Larkin of St. Louis, taking a trip over the Trail in hopes of curing his ailments, jotted in his "memorandum book" on October 6, 1856: "Wolves–white & gray–ranging about the prairies in plenty."[10] Goldseeker David Kellogg, on his way to the goldfields in Colorado, recorded in his diary on September 30, 1858, at Cow Creek: "Caught some fish in this stream. The country well watered and grass luxuriant. Buffalo and their attendants, big white wolves, very plentiful.[11] Earlier, Garrard recalled the first white wolf he saw as he and several other men from the caravan were heading away from their camp grounds to hunt buffalo: "I had a shot at a large white wolf—the first we had seen. He was loping around the camp when my shrill whistle brought him to a stand still; at the sharp crack of my rifle, the keen-faced stranger took to the plain."[12]

Some chroniclers, including Gregg, thought the white wolves were probably very old animals, but that did not seem likely to others. It was also thought by those who observed white wolves that they were not albinos. As time passed and the Trail became more heavily traveled, only gray wolves were mentioned in the written records. The white wolves seem to have disappeared. No explanation for this occurrence was uncovered in biological or historical writings. Robert H. Busch briefly states about the white wolf in his book *The Wolf Almanac* (1995): "Today, white wolves have been reported as far south as Minnesota, but in the past they appear to have been quite common in the plains. The Lewis and Clark expedition reported large numbers of almost-white wolves on the plains. . . . Painter George Catlin wrote in 1841 of his travels across the American plains 'where the wolves are white.'"[13] There are still some areas in North America, such as northern Canada, where white wolves are found.

———

Frank S. Edwards wrote in his book *A Campaign in New Mexico with Colonel Doniphan* (1847) that wolves would not attack humans. Some people taking the Trail agreed with him. Edwards wrote: "They are generally seen in packs, and will scent fresh meat or blood at a great distance (as much as a mile or more), and being exceedingly cowardly they never attack man—and unless driven by hunger will not kill any animal, preferring dead carcasses."[14] Gregg expressed a similar opinion in *Commerce of the Prairies*: "I have never known these animals, rapacious as they are, [to] extend their attacks to man, though they probably would if very hungry and a favorable opportunity presented itself."[15]

A few men, however, told of very close encounters with wolves. One encounter happened to Captain James McClure as he walked ahead of his supply train, only to be chased back by wolves. "I will never forget the terrible ordeal," he later declared, "and the relief I felt when I found myself safe from their fangs."[16] The wolf's four fangs at the front of the mouth may be as long as two inches from root to tip and are specialized for grabbing, wounding, and killing prey. The other thirty-eight teeth perform other specific functions, including sharp side teeth to cut tough muscle and flat teeth in the back of the mouth to crush bone so that it can be swallowed easily.

Despite his earlier statement, Gregg did have an encounter with a gray wolf, although it was not as threatening as McClure's, and Gregg admitted that he had been the one to make the first challenge: "I shall not soon forget an adventure with one of them, many years ago, on the frontier of Missouri. Riding near the prairie border, I perceived one of the largest and fiercest of the gray species, which had just descended from the west, and seemed famished to desperation. I at once prepared for a chase; and, being without arms, I caught up a cudgel, when I betook me valiantly to the charge, much stronger, as I soon discovered, in my cause than in my equipment. The wolf was in no humor to flee, however, but boldly met me full half-way. I was soon disarmed, for my club broke upon the animal's head. He then 'laid to' my horse's legs, which, not relishing the conflict, gave a plunge and sent me whirling over his head, and made his escape, leaving me and the wolf at close quarters.

"I was no sooner upon my feet than my antagonist renewed the charge; but, being without weapon, or any means of awakening an emotion of terror, save through his imagination, I took off my large black hat, used it for a shield, and began to thrust it towards his gaping jaws. My *ruse* had the desired effect; for, after springing at me a few times, he wheeled about and trotted off several paces, and stopped to gaze at me. Being apprehensive that he might change his mind and return to the attack, and conscious that, under the compromise, I had the best of the bargain, I very resolutely—took to my heels, glad of the opportunity of making a drawn game, though I had myself given the challenge."[17]

Richens Lacy "Uncle Dick" Wootton, mountain man and builder-operator of a toll road over Raton Pass on the Mountain Route, detested the wolves who pestered him, although he did not fear they would attack him. He recalled the annoyance they caused: "When hunting buffalo, I have sat many a time all night by a blazing fire, throwing the red-hot brands every now and then at a pack of wolves, to keep them from stealing the game which I had slaughtered."[18]

Sometimes the aggravation of the wolves' company proved to be too much for Wootton: "There must have been hundreds of the vicious brutes in the pack that kept me company all night. They would come so close that I could see their eyes shining like balls of fire in the darkness, and all the time they kept up a snapping and snarling which would have set a man crazy who did not know what cowardly brutes they were." Declaring that he did not usually waste ammunition to kill wolves, he continued: "I killed three or four during the night and the dead wolves were at once torn to pieces and devoured by the balance of the dirty gang of cannibals. They sneaked away just before daylight came in the morning, but they had given me a might lively all-night serenade."[19]

An incident involving a wolf in the 1860s was told by Theodore Weichselbaum in his recollections of experiences as a merchant on the Trail: "In one of my contracts out at Fort Larned, I hired [James M.] Harvey and his ox team. He was with me thirty days on the trip. I saved his life near Larned. A large white wolf frothing at the mouth had attacked him when I happened to be near. I drew my revolver and killed the wolf."[20] Years later, Harvey became the governor of Kansas.

The attacking wolf was probably rabid if it was "frothing at the mouth," and most people on the prairies and plains knew that a rabid animal, whether wild or domesticated, was apt to attack and kill a human being—or inflict a wound that would cause the injured person to suffer a horrible, painful death from hydrophobia (rabies). Such an incident occurred at Fort Larned in August 1868 when a rabid wolf invaded the post and attacked several people, including Corporal Mike McGuillicuddy, who was a patient in the post hospital. The other victims were treated by the post surgeon and survived, but Corporal McGuillicuddy refused to have a torn finger amputated, and he died from hydrophobia a month later. A dog at the post was also bitten by the wolf and died from the disease.[21]

Considering the overwhelming numbers of wolves, it is worthy of notice that they seldom threatened the lives of travelers on the Trail. Nevertheless, wolves could be very troublesome, adding to the numerous annoyances people faced along its routes. Hezekiah Brake wrote of such in a narrative of his experiences when he took the Trail in 1858 to the Fort Union area. "We crossed the Cimarron [River] that night and drank a cup of tea on the opposite bank. Wrapped in our blankets, we lay down as usual to sleep, but something kept me awake: I did not know but what it might be a prairie dog or antelope [his dinner]. Louis [a member of his party] was the sleepiest of mortals. Once asleep, nothing short of an earthquake would have disturbed him.

"As I slept lightly and wakened easily, I always kept my boots and my only pony bridle under my head, in order that I might be prepared for any emergency that might arise. I had just fallen asleep, when I felt something move under my head. I put up my hand [and discovered] one of the boots and the bridle were gone. I sprang up in time to see in the dim light, the outline of a large wolf, but the yell I gave must have disconcerted his wolf-ship, for he ran, leaving the boot and bridle. My companion [Louis] knew nothing of it the next morning, and but for the condition of my property, would have kept on insisting that it was all 'a bootless dream.' I had no fancy, however, to ride into Fort Union on a pony wearing a rope bridle, myself minus one boot, and I praised the Fates that I had recovered my confiscated goods."[22]

British adventurer George Frederick Augustus Ruxton chronicled his travels from Mexico through El Paso, across New Mexico, and into Colorado in his celebrated book *Adventures in Mexico and the Rocky Mountains* (1847). Continually amazed at the brazen boldness and sagacity of wolves and coyotes, he recalled camping along a stream in a thickly timbered bottom where these animals were found in large numbers: "I could scarcely leave my saddles for a few minutes on the ground without finding the straps of raw-hide gnawed to pieces; and one night the hungry brutes ate up all the ropes which were tied on the necks of the animals and trailed along the ground. [The ropes] were actually devoured to within a yard of the mules' throats."[23]

In *Life in the Far West* (1849), Ruxton wrote about mountain men and the fur trade, and included more about wolves' behaviors: "Wolves are so common on the plains and in the mountains that the hunter never cares to throw away a charge of ammunition upon them, although the ravenous animals are a constant source of annoyance to him, creeping to the camp-fire at night and gnawing his saddles and apishamores [saddle blankets made of buffalo-calf skin], eating the skin ropes which secure the horses and mules to their pickets, and even their very hobbles, and not unfrequently killing or entirely disabling the animals themselves."[24]

On his way to Colorado, Ruxton and his small party were followed for days by a lone wolf—"a large grey wolf." He recounted this experience of having an unusual follower: "Every evening, as soon as we got into camp, he made his appearance, squatting quietly at a little distance, and after we had turned in for the night helping himself to anything lying about. Our first acquaintance commenced on the prairie when I had killed two ante-lope, and the excellent dinner he then made, on the remains of the two carcasses, had evidently attached him to our society. In the morning, as soon as we left the camp, he took possession, and quickly ate up the remnants of our supper and some little extras I always took care to leave for him.

"Shortly after, he would trot after us and if we halted for a short time to adjust the mule-packs or water the animals, he sat down quietly until we resumed our march. But when I killed an antelope, and was in the act of butchering it, he gravely looked on, or loped round and round, licking his jaws, and in a state of evident self-gratulation. I had him twenty times a day within reach of my rifle, but he became such an old friend that I never dreamed of molesting him."[25] A century later, Southwestern folklorist and author J. Frank Dobie commented on Ruxton's sympathy toward the lone wolf: "No American contemporary of Ruxton's on the frontier would have resisted killing that wolf."[26]

In his reminiscences of his Trail days, W. B. Napton described in *Over the Santa Fé Trail, 1857,* a wolf chase that was indelibly impressed upon his memory: "Riding thus alone on one occasion, some distance ahead of the train, I saw a large gray wolf galloping across my course, going towards the road [Santa Fe Trail]. I determined to give him chase, and after him I went. The wolf increased his speed, and urging my horse to his best, we went flying across the road one hundred yards in front of the train and in full view of it.

"As we flew by, the entire company of teamsters gave us an encouraging whoop, but whether designed for me or the wolf I was not able to determine. I had followed the big fellow closely for a mile, emptying at him, if not in him, the entire twelve chambers of my revolvers. At one time within twenty feet of him, but not having any ammunition for reloading with me, nor time for recharging my pistols if I had, he disappeared over the ridge and I saw nothing more of him."[27]

The howling of wolves was a frequent topic in narratives of the Trail's travelers. Their howl has a long, mournful slide from a high note to one of an octave or more lower. James Ross Larkin was among the number who recorded their dislike of the sound: "Wolves howling & crying near our camp make a hideous noise."[28] The adjective "hideous" seems to have been the word of choice among those who described the howling. It caused many sleepless hours and nights, at least until the fatigue from the long day's journey, sometimes extending until late after dark, inured people to the nightly "serenades."

Frank S. Edwards told of an ear-offending incident caused by a wolf while he was standing sentinel on the outer edge of his company's horses at Big Timbers on the Arkansas: "I was leaning upon my carbine, with my back to a small ravine along the edge of which my post extended and my mind in a quiet reverie, when, suddenly, from behind a bush, not three feet from me, a big gray wolf set up his dismal cry unconscious of my presence. It, annoyingly, took me by surprise. Snatching up a stone, I hurled it after his howling wolfship as he dashed precipitately down the ravine. I would have given something to have been allowed to shoot him, but as orders were to shoot nothing of less size than an Indian, I dared not alarm the camp by a shot."[29] It was customary for Trail companies to have a regulation that no one fire a gun in or around camp unless one's life or the safety of the company depended on it.

Wolves' pelts were valuable in the fur trade, and most wolf hunts were conducted by people who made a living in that trade. James Josiah Webb, a prominent merchant engaged in the trade with New Mexico, recorded his recollections about the hunting of wolves: "To give some idea of the numbers of wolves on the prairie in the buffalo range, I will give an account of two men formerly conductors of the mail from Independence to Santa Fe. I think it was in 1854 or 1855 when they went to Walnut Creek and built a small mud fort, and in summer they would sell what few knickknacks they could to traders and passing travelers, and in winter their business was to kill wolves for the pelts. They would kill a buffalo and cut the meat in small pieces and scatter it about in all directions a half a mile or so from camp, and so bait the wolves for about two days. Meantime, all hands were preparing meat in pieces about two inches square, cutting a slit in the middle and opening it and putting a quantity of strychnine in the center and closing the parts upon it. When a sufficient amount was prepared, and the wolves were well baited, they would put out the poisoned meat. One morning after putting out the poison, they picked up sixty-four wolves, and none of them over a mile and a half from camp. The proceeds from that winter's hunt were over four thousand dollars."[30]

As years passed and the numbers of buffalo and other native prey declined sharply, wolves turned to livestock on the farms and ranches of the frontier for their food supply. The "scepter of authority," once held so firmly by wolves, was now in the hands of another intimidating predator— one with two legs. The shooting, trapping, poisoning, and "denning" (removal and killing of wolf pups still in their dens) by government and private hunters increased with a vengeance during the 1880s. The native gray wolf was soon gone from the plains and mountain region of the old Trail.

The campaign to eradicate wolves culminated in the "Great Wolf Slaughter" in New Mexico, southern Arizona, and West Texas between 1890 and 1920. By 1930, gray wolves had nearly been eliminated in New Mexico, across the Southwest and West, and the lower 48. By 1970, the gray wolf had been extirpated over most of this animal's former range in the United States and Mexico.[31] Today, their range includes parts of Alaska, northern Canada, and the northern Rocky Mountains.

The Mexican gray wolf, or El Lobo, a sub-species (*Canis lupus baileyi*) and about half the size of its cousin (*Canis lupus*) was able to hang on longer in Mexico, and sometimes strayed north from Mexico. The U.S. federal government designated the Mexican gray wolf as an endangered species in 1976. None were thought to survive in the wild and only about thirty existed in captivity.[32] Captive breeding programs, begun in the 1980s in Arizona and New Mexico, proved successful, and interest grew over the following years in the development of reintroduction and recovery programs that would restore a small number of wolves to selected wilderness areas in their historic range. In the late 1990s, captive-bred Mexican wolves were released in the Apache National Wilderness of eastern Arizona, which is adjacent to the Gila National Wilderness in western New Mexico.

Although relocation of wild-trapped wolves from Canada into the Yellowstone National Park ecosystem in 1995 has had considerable success, the reintroduction of captive-raised Mexican wolves into the Southwest, begun in 1998, has been fraught with numerous problems. Political and economic issues, controversy over recommended changes in program management, illegal poaching, wanton killing, and other problems have contributed to less than encouraging results. Research has shown that wolves play an important role in keeping their ecosystem healthy. It also emphasizes the importance to animals of the wilderness, such as wolves, grizzlies, and wolverines, of having large areas of contiguous wild habitat that can provide them with enough food or prey and protected areas far off the beaten roads and trails for them to thrive and raise their offspring. Native habitat is a necessity.

We know that long-standing hatreds die hard, particularly when one's livelihood and security seem threatened. This is certainly true for the hatred of wolves, which goes back long before the days of the Santa Fe Trail. Although there are wolves now roaming in some wilderness areas, their future remains in doubt. In much of their historic range in North America, the howling of wolves is a sound lost to the passage of time.

Chapter Five

Coyotes and Roadrunners

Coyotes, like wolves, are wild members of the dog family Canidae. The coyote, however, is smaller and has a smaller nose pad and larger ears than the wolf. When running, the coyote usually holds his tail lower, or down, while the wolf holds his in a more horizontal position. Frequently called the prairie wolf by frontiersmen, this canid was also known as the coyote wolf, brush wolf, or prairie jackal. When explorers Meriwether Lewis and William Clark first saw a coyote, they thought he was a fox. Soon realizing the difference between the two, they called the coyote a "prairie wolf." The name "coyote" is derived from "*coyótl,*" a Nahuatl word of the Aztecs of Mexico. A description of the *coyótl* is found in the first volume of Francesco Saverio Clavigero's *Historia Antigua de Méjico,* published in Mexico City in 1844. It describes the coyote as "similar to the wolf in voracity, to the fox in cunning, to the dog in shape, and in other propensities to the *adive,* or *jackal;* for which reason some Mexican writers have counted it among several of these species; but it undoubtedly is different from all of them."[1]

Coyotes did their share of howling along the Trail, to the annoyance of some travelers and to the enjoyment of others. Their scientific name, *Canis latrans,* literally means "barking dog." Josiah Gregg declared that this designation was "a distinction to which its noisiness emphatically entitles it."[2] Marion Sloan Russell remembered hearing the howling of coyotes when she was a young girl on the Trail, traveling with her mother and brother, Will, in the early 1850s. She recalled the sound many years later in her memoirs, *Land of Enchantment:* "While most of the drivers slept under the wagons, the women and children slept inside the wagons or in tents. Each night we pitched the tent close to the wagon and it spread its dark wings over the three of us. . . . I would awaken to hear the coyote's eerie cry in the

darkness. I would creep close to mother and shiver. Sometimes one of the mules would start braying, and others would take it up, making the night hideous."[3]

The howl of the coyote, heard particularly at dawn and dusk, was considered more melodious by adventurer Lewis Garrard. He described their "song" in *Wah-to-yah and the Taos Trail:* "For the first night or two after entering the buffalo region, we were serenaded by the coyote wolf, a species of music much like a commingled bark, whine, yelp, and occasionally a spasmodic laugh, now tenor, now basso; then one would take a treble solo, and after an ear-piercing prelude, all would join in chorus, making an indescribable discord."[4]

Their variety of tones and range of an octave or more, rising and falling, can make the howling of two coyotes "singing" together sound like a pack. Gregg thought they outdid wolves in being noisy: "Like ventriloquists, a pair of them will represent a dozen distinct voices in such succession—will bark, chatter, yelp, whine, and howl in such a variety of note, that one would fancy a score of them at hand."[5] This is an illusion especially common with coyotes, who punctuate their howls with lots of yapping and yipping. Coyotes can also bark, but do it much less than dogs, their domesticated kin.

Among others who found their song appealing was Lieutenant Philip St. George Cooke, assigned to the first military escort accompanying the trading caravans over the Trail during the summer and fall of 1829. He remarked in his journal: "I never heard without pleasure the voice of the Night."[6] Years later in the 1910s, when the federal government and others were waging on all fronts a war of extermination against coyotes, naturalist Ernest Thompson Seton wrote in his *Wild Animals at Home* (1917): "I must confess that if by any means they should succeed in exterminating the Coyote in the West, I should feel that I had lost something of very great value. I never fail to get that joyful thrill when the 'Medicine Dogs' sing their 'Medicine Song' in the dusk, or the equally weird and thrilling chorus with which they greet the dawn; for they have a large repertoire and a remarkable register."[7]

Researchers studying coyotes believe they howl for several reasons: to communicate with other coyotes; to announce their location and presence in their territory; to reinforce bonds; and to announce a change in the weather. Research has pointed to a falling barometer as a cause of howling. Some observers are convinced that coyotes also howl for the pure enjoyment they derive from it.

———

Although able to compete with the large gray wolf in howling, the coyote is smaller and less powerful. Adult males are about four feet long (including a tail of about fourteen inches), stand two feet at the shoulder, and weigh twenty to twenty-five pounds.[8] Most live in pairs or alone, although they may form a pack of three or more members. The number in a pack is largely a function of carrying capacity, or food supply, of their habitat.

On the buffalo range during Trail days, they fed on the remains of carcasses left by wolves and men. They preyed on pronghorn, goats, sheep, and cattle, in particular the young or infirm, and on rabbits and rodents, including prairie dogs, gophers, mice, and rats. They also eat a variety of reptiles and insects. Well-known for their taste for chicken, other fowl or birds may do for a meal. They will eat what is available in their environments or in season, such as juniper berries, mesquite beans, and watermelons. J. Frank Dobie, a collector of Southwest lore, once remarked: "The coyote's favorite food is anything he can chew; it does not have to be digestible."[9]

James Brice, a mail carrier on the Santa Fe Trail in the 1860s, once saw a few coyotes bring down a buffalo: "When I was conductor running with the mail, I saw three coyotes attack a lone buffalo on the Arkansas, opposite Fort Mann. One kept jumping at the buffalo's head as if trying to catch his nose and two kept jumping at his hamstrings until they severed them. Then [the buffalo] became powerless, falling down, and they pounced upon his body and began to devour him."[10]

British adventurer George Frederick Augustus Ruxton recorded his impressions of "the *coyote* or the *cayeute*" of the mountaineers, the "wach-unkamnet" or "medicine wolf" of the Indians. He observed them taking turns running down large prey: "This little wolf, whose fur is of great thickness and beauty, although of diminutive size, is wonderfully sagacious, and makes up by cunning what it wants in physical strength. In bands of from three to thirty, they will not unfrequently station themselves along the 'runs' of the deer and antelope, extending their line for many miles—and the quarry being started, each will follow in pursuit until tired, when it relinquishes the chase to another relay, following slowly after until the animal is fairly run down, when all hurry to the spot and speedily consume the carcass."[11] Other adventurers observed those relays to catch game.

Ruxton added: "The *cayeute,* however, is often made a tool of by his larger brethren, unless indeed he acts from motives of spontaneous charity. When a hunter has slaughtered game and is in the act of butchering it, [coyotes] sit patiently at a short distance from the scene of operations, while at a more respectful one, the large wolves (the white or gray) lope hungrily around, licking their chops in hungry expectation. Not unfrequently the hunter throws a piece of meat towards the smaller one, who seizes it immediately, and runs off with the morsel in his mouth. Before he gets many yards with his prize, the large wolf pounces with a growl upon him, and the *cayeute,* dropping the meat, returns to his former position, and will continue his charitable act as long as the hunter pleases to supply him."[12] When meat was available, the wolf, capable of killing a coyote with one crush of his jaws, always ate first.

In his vivid description of Bent's Fort on the Mountain Route of the Trail, which he visited in the spring of 1847, Ruxton noted the presence of coyotes: "Outside the fort, at any hour of the day or night, one may safely wager to see a dozen coyotes, or prairie wolves, loping round or seated on their haunches, and looking gravely on, waiting patiently for some chance offal to be cast outside."[13]

Stanley Vestal commented in his popular book *The Old Santa Fe Trail* that coyotes also gave "good sport" on the Trail for "the hunters who would ride a little ahead of the caravan, with their hounds at heel." He elaborated: "In chasing coyotes the hunter had to ride hard, that animal not being given to silly tricks like waiting on his enemies. He was gone to cover, and generally quite literally 'gone away,' before the horsemen came near enough to see anything of the chase."[14] Vestal also remarked that this "sport" was too hard on their horses during the hot summer. Those chases could also be very hard on their dogs. Some fell to the ground from heat exhaustion and died.

Once living only on the plains of western North America, coyotes now live throughout most of the United States. Their range, unlike the ranges of many animals of the prairies and plains, has significantly increased. In fact, they have expanded their range threefold since the 1850s.[15] Their adaptability and resilience have been important factors in their survival and the expansion of their range under seemingly impossible odds. They live in many types of habitat today—deserts, plains, mountains, woodlands, agricultural lands, urban and suburban areas, industrial areas, parks, and golf courses—whatever suits their needs in terms of food and shelter.

———

While many travelers loathed wolves with a vengeance and a number were not particularly fond of coyotes (although Stanley Vestal had the opinion that "most men on the Trail had a sneaking fondness for the coyote"),[16] almost all of the people were delighted to encounter roadrunners. No other bird looks quite like the roadrunner. Considered a symbol of good luck, these fascinating birds were frequently seen on the Trail, especially in New Mexico, where today they are the official state bird. Their range extends west across southwestern United States from Texas to California and in southern Kansas and parts of Colorado and Oklahoma. They also range south of the border as far as central Mexico. Although found in a few other states, ground-dwelling roadrunners prefer regions with lots of sunshine, low humidity, and scrub, or desert, vegetation, such as chaparral and mesquite.

Roadrunners got their name from running down trails and roads ahead of horses and wagons, seeming to challenge those behind them to a race, and then darting like a flash into the brush. Marion Sloan Russell recalled seeing these fascinating birds along the Trail: "Birds with long tails would walk the trail before us; walk upright and faster than our mules could walk. The drivers called them road-runners."[17] The people on the Trail especially enjoyed these colorful birds' amusing behavior, hunting antics, and curiosity about human activities.

The subject of legend and lore for centuries, the Greater roadrunner (*Geococcyx californianus*) is a ground-dwelling bird belonging to the cuckoo family Cuculidae. The roadrunner has been given a variety of names over the years, varying according to locality: the chaparral or chaparral cock; the ground cuckoo; snake killer or lizard killer; the prairie pheasant; *paisano* (Spanish for fellow countryman or compatriot); and *corre camino* (Spanish for "runs the road"). Among these and other names, *paisano* has been one of the most popular and is still commonly used in parts of the Southwest and in Mexico.

On January 11, 1847, Lieutenant James William Abert was returning east on the Trail from Santa Fe and headed to Fort Leavenworth in Kansas, after completing his observations for a scientific report about New Mexico. He wrote in his journal: "Today we saw some curious birds, which our old hunters called the *paisano*."[18] Since so little information was uncovered about these birds, Abert included some provided by another person, much of which was inaccurate. He added at the end of his entry: "This bird is found throughout the Raton Pass [on the Mountain Route], and some individual specimens have been seen on the Arkansas river, a few miles to the west of Bent's Fort."[19]

Roadrunners measure about two feet long, and half of their length is tail feathers, which assist them in very rapid braking. Unlike their cuckoo cousins, they do not fly long distances, although they glide very well. Sometimes they will fly to a lower branch on a tree, a shrub, or a fence post from which they can view the ground more easily. Also unlike some of their cousins, who are known to be parasitic nesters, laying their eggs in other birds' nests, roadrunners build their own nests, hatch their own young, and are attentive, all-around good parents to their offspring.

Roadrunners excel on the ground. Long, sturdy legs, a streamlined body, and a strong tail used as a rudder give them the ability to maintain fifteen miles per hour for a running distance of three hundred yards. Walking stride measures six to eight inches, lengthening to twenty inches when running, with head, body, and tail stretched out almost as straight as an arrow.[20] Like other members of the cuckoo family, their feet are zygodactyl, meaning two toes point forward and two point backward. Their X-shaped feet provide stability in walking and running. Their tracks, which look like an X in sand or dirt, make it impossible to tell whether they are coming or going. This particular characteristic has greatly added to their mystique over the centuries.

Since long before the days of the Santa Fe Trail, some Indians of the Southwest have considered the roadrunner a bird of great magic and supernatural powers, a "medicine bird" or "war bird," symbolizing bravery, strength, and endurance. Their X-shaped track has been used in a variety of ways, including body decoration, in warrior or hunting ceremonies, on warriors' shields, and as a mark of protection to keep away evil spirits and enemies. In some Indian cultures, for instance, a secure afterlife is guaranteed if the roadrunner's tracks are placed around the house of a deceased person. The X-mark causes confusion for evil spirits, because the direction taken by a departed soul cannot be ascertained. The roadrunner's feathers have also been used in various ways for protection, such as placing them on a baby's cradleboard to keep away evil spirits.

Stories have been told for centuries by the Indians and, in more recent times, by Mexicans, frontiersmen, and settlers, of the roadrunner's courage and ability to attack and kill rattlesnakes. For years, those stories were not believed by scientists until they witnessed for themselves the skill used by roadrunners, sometimes working in pairs, to kill a rattlesnake. When attacking a rattlesnake, the roadrunner crouches low, circling the snake and dropping his wings to test the snake's quickness. When the snake strikes, the roadrunner leaps out of the way of its fangs, and then immediately leaps forward, using his long, strong beak to strike the snake's head, the most vulnerable part of this reptile. In a flurry of activity, the snake is killed and swallowed headfirst. The roadrunner seldom attacks a snake over two feet long. A rattlesnake's venom and a scorpion's poison do not affect a roadrunner's digestive system. Sometimes part of a snake's or lizard's tail is seen hanging out of a roadrunner's mouth, because the reptile cannot be swallowed whole and has to be digested over a period of time, usually a matter of hours.[21]

The largest part of a roadrunner's diet consists of grasshoppers, crickets, and other insects. In addition to insects and reptiles, this bird also eats mice, rats, and other small rodents. When a rodent is too big to swallow whole, the roadrunner beats it on a rock or hard surface, making it softer and easier to swallow. Gregg mentioned in his description of wild birds in *Commerce of the Praries* that roadrunners are good mousers: "There is to be found in Chihuahua and other southern districts a very beautiful bird called *paisano* (literally 'countryman'), which when domesticated, performs all the offices of a cat in ridding the dwelling-houses of mice and other vermin."[22]

Stories have been told about roadrunners attaching themselves to people, and some country folk consider them easy to domesticate, especially at an early age. Marc Simmons wrote about this in one of his history columns: "One case is known of a roadrunner befriending a farmer and following him every day as he plowed his fields with oxen. Perhaps this friendly trait explains why the Spanish-speaking folk called the roadrunner *paisano*."[23] Of course, this amazing bird is doing what he does best and that is hunting and eating insects. Whether carefully examining a place or following a person, plow, tractor, or another creature, the roadrunner is constantly on the lookout for a snack or meal, depending upon whatever is scared up along the way.

During the 1930s, a bounty was placed on roadrunners because they were charged by hunters of eating quail's eggs and their young ones. This rapidly led to the hunting down and killing on sight of large numbers of roadrunners. Subsequent studies of the stomach contents of dead roadrunners proved that they are not destroyers of quails or quail's eggs. It was found that, on rare occasions, they will eat a small bird. But given that the bulk of their food consists of insects, reptiles, and small rodents, they are a valuable helper to farmers and ranchers. J. Frank Dobie, friend of roadrunners and collector of lore and memorabilia about them, declared this unfortunate episode in the roadrunners' history "unjust persecution."[24] Today, roadrunners are protected by state and federal laws.

An observer of roadrunners for many years, Dobie wrote about the fascinating behavior and characteristics of the *paisano*: "The bird has a great deal of curiosity. . . . One will hop into the open door of a house and stand there a long time, looking this way and that. Perhaps he has an idea that some shade-loving creature suited to his diet is in the house. He will come up to a camp to investigate in the same way. I never tire of watching one of these birds dart down a trail or road, suddenly throw on the brakes by hoisting his tail, stand for a minute dead still except for panting and cocking his head to one side and then to the other, and then suddenly streak out again. The way he raises and lowers the plumage on his lustrous-feathered head while he goes crut, crut, crut with his vocal organs is an endless fascination. He must surely be the most comical bird of America."[25]

———

The roadrunner and coyote have become closely linked in the minds of many people today through modern-day media, although some are unaware that the roadrunner is a real, living bird, and not one created by storytellers. Simmons, also an admirer of roadrunners, has remarked about this: "Strangers to the Southwest are nearly always amazed to learn that there really is such a feathered creature as the roadrunner. Apparently, the movie cartoons about a running bird that outwits the coyote have led many people to believe that the whole thing is a joke invented by the filmmakers."[26]

Simmons, of course, is referring to Wile E. Coyote (*Eatimus anythingus*) and the Road Runner (*Accelerati incredibulus*) of "silver screen" fame. Created by the late animation director Chuck Jones and writer Michael Maltese, their first fast-paced adventure, "Fast and Furry-ous," appeared in 1949. It set the pace for more than twenty years of episodes, followed by years of television reruns, videotape replays, and video games. Each adventure faithfully held to the same formula—a classic ritual, as popular today as the old tales told around campfires for centuries, of the most determined predator in pursuit of the fastest prey.

Although some may believe that roadrunners actually make the sound "beep-beep," author-photographer Wyman Meinzer states with certainty that they definitely do not "beep." In fact, he is able to distinguish at least sixteen different vocalizations or sounds made by them, including a soft cooing, a whine, a bark similar to the coyote's yip, a cluck, a barely audible hoot, a whir, a hum, and a clacking or clattering sound made by popping together the upper and lower parts of their beaks.[27]

Today, roadrunners and coyotes are considered living symbols of the desert. Their images are found everywhere across the American Southwest and West, even in other parts of the country. They are seen on billboards and neon signs, in advertising, and as logos. They are also subjects of popular arts and crafts. Many types of businesses and services use "roadrunner" and "coyote" in their company or restaurant names, such as the famous Coyote Cafe in Santa Fe, located close to the End of the Trail monument on the southeast corner of the Santa Fe Plaza. Such widespread popularity may cause one to wonder what the people on the Santa Fe Trail long ago might have thought of coyotes' and roadrunners' far-reaching fame.

Chapter Six

Prairie Chicken

On a bone-chilling April morning long before sunrise, a small group of men and women huddle in a circle close to their modern "wagon," which has brought them to private ranch land near the old Cimarron Route of the Santa Fe National Historic Trail. In this vast grassland region are 108,000 acres designated as the Cimarron National Grassland, the largest area of public land in Kansas. During the time of the Santa Fe Trail, the wagons of traders, merchants, and freighters, as well as a multitude of travelers in a variety of conveyances or on horseback were headed over the land on their way to Santa Fe, while others were sometimes headed eastward to Missouri. The Cimarron National Grassland today includes twenty-three miles of the Santa Fe National Historic Trail. There is a nineteen-mile companion trail, very close to the old Trail, that hikers follow.

This morning's travelers, bundled from head to toe, are serious bird-watchers. Their common goal is not to reach Santa Fe, but to witness a rite of spring described by an observer as "one of the most spectacular sights in natural history."[1] Like the Trail travelers long ago, these people have brought along some grub and gear. The grub includes a cup of hot coffee and something a bit similar to a griddle cake but with a hole in the center. Their gear includes cameras, spotting scopes, and binoculars. Once aboard their "wagon," they follow a narrow dirt road in the darkness, reaching their destination on the grassland about an hour before sunrise. They will have access to a small, simple structure that serves as a viewing or observation blind and will keep them out of sight. They make themselves as comfortable as possible to quietly await a great performance that begins about half an hour before sunrise—the annual courtship ritual of the lesser prairie chicken.

"Prairie chicken" is the common name of two species of North American grouse (family Tetraonidae): the lesser prairie chicken (*Tympanuchus pallidicinctus*), which was the species found "in great abundance" by travelers on the Trail as they crossed the great expanse of prairies on their way southwest and west; and the greater prairie chicken (*Tympanuchus cupido*).[2] Ground-dwelling, chickenlike game birds, they are larger than quail and feed on seeds, leaves, insects, buds, berries, and grains. Closely related and similar in appearance, both species have predominantly brown and white barred feathers, a rounded darker tail, a black bar across the eye, and a small, yellow-orange brow comb. Physical differences are primarily in their size and the color of the air sacs, called tympani, on either side of the throat. Courtship habits and rituals are also similar. However, their prairie habitats and ranges differ.[3]

Lesser prairie chicken measure about sixteen inches long, have a wingspread of about twenty-five to twenty-six inches, weigh less than two pounds, and have deep reddish-colored air sacs.[4] Their habitat is sand-sage prairie or sandhill areas with short- and mid-length grasses, sagebrush, and shrub-like plants, which provide food, cover, and nesting places. Their range is mostly southwest of the range of the greater prairie chicken and largely in southwestern Kansas, extreme southeastern Colorado, eastern New Mexico, and the Panhandle country of Oklahoma and Texas.[5] Today it is believed they occupy only about 10 percent of their historical range in North America.[6] Cimarron National Grassland has supported a portion of the remaining population and has been considered one of the premier spots to observe these fascinating birds. Those living on the Cimarron National Grassland have generally been found south of the Cimarron River on a strip about two to five miles wide.[7] With lingering dry periods, which reduce vegetation for cover and provide less food for them and their broods, the population count has fluctuated.

Greater prairie chicken are approximately eighteen inches long with a wingspread of twenty-eight inches, weigh about two and one-half pounds, and have yellow-orange air sacs.[8] These birds prefer native tall-grass prairies and have a range extending from North Dakota and Minnesota south through the Great Plains and into Texas. Their range may at times overlap that of the lesser prairie chicken. They are more common in Kansas, inhabiting the central and eastern parts of the state. The Flint Hills of Kansas, the last remaining large expanse of tall-grass prairie, is one of the places where these birds may be observed. With the disappearance of tall-grass prairie from many areas, greater prairie chicken have been able to adapt to some extent to pasture and cropland, if good cover is present for raising their young. Cropland also may provide additional food, particularly in the winter months.

These grouse of the prairies constitute one of four distinct families of land birds, which also include turkeys, quail, partridges, and pheasants, that were plentiful along parts of the Trail. All have well-developed legs for walking and running. They have three long, forward-pointing toes on each foot, used for scratching the ground, and a small hind toe.[9] They are able to fly and can burst into full flight with rapid wingbeats from a sitting position. The flight of a prairie chicken is considered a thrilling sight by many observers. It has been described as "a brief burst of rapid pulses from the short wings, followed by a long fixed-wing glide."[10] This is known as "volplaning," a term normally used in aviation referring to their bent-down-wing glide. Author William Least Heat-Moon saw them burst into the air while hiking across grassland in Chase County, Kansas, and later wrote in his book *Prairy-Erth:* "Prairie chicken broke noisily and did their sweet dihedral-winged glides to new cover."[11] Their flight can be deceiving, because they are able to exceed thirty miles per hour,[12] although they usually do not go farther than a few hundred feet.

———

Back at the observation site, nearly an hour has passed and first light has broken. Having waited patiently in place, the avid bird-watchers stir in anticipation. The male lesser prairie chicken begin arriving on their leks, the display areas where the courtship ritual takes place. Leks are also called "booming grounds," a term that has generally been used for the leks of both lesser and greater prairie chicken. These are located on rises or slightly elevated areas where vegetation is generally sparser and the visibility is better than on lower ground. The prime time for the observation of their rite of spring is from mid-March through May, with peak activity generally occurring in April.

The males, or cocks, take up their positions on the territories they claimed earlier, after their arrival on the leks in late February or early March. The courtship display of male lesser prairie chicken commences with deep bowing, the dropping of wings at their sides, the spreading and raising of the rounded tail, and the raising of black neck feathers, called pinnae feathers, which usually lie flat, but when raised make them look as though they have horns. Their fancy footwork is composed of short steps and rapid foot stamping. The colorful air sacs are fully inflated with air and a booming sound is emitted upon release of the air that resembles "a hollow oo-loo-woo," followed by cackling and clucking."[13] The booming has also been described as the combined sounds of timpani drum and empty, blow-across-the-top Coca-Cola bottle. Their calls are higher-pitched and shorter than that of greater prairie chicken, which have a longer, low-pitched, hooting moan, also followed by clucking. The stillness of the prairie is filled with wondrous sounds, and most observers are not disappointed. On a quiet, calm day their booms may be heard up to a mile away.

The hens begin congregating after the booming has started. They watch the males' displays from the cover on the edges of the leks. They visit the leks where males are performing for about a two- or three-week period, and during this time will eventually select a mate. In fierce rivalry for a mate, a male will sometimes leap into the air and confront other males, but they do not actually fight. This is more a show of bravado to keep the other males from encroaching on his territory. As a hen approaches a booming male, usually one of those with a "preferred" territory in the central part of a lek, he prostrates himself on the ground. The hen steps up to him, bowing and slightly spreading her wings, to show that he is "the chosen one." Following mating, the male returns to the lek to continue the ritual and seek another mate, while the hen departs to attend to the tasks of nesting and raising her brood. She will lay ten to fourteen eggs, which have an incubation period of about twenty-three or twenty-four days.[14] The males generally leave the leks at the end of spring and disperse to their habitats on the grassland. They usually return to the same booming ground, or lek, year after year.

———

An earlier observer of prairie chicken was Marion Sloan Russell. She vividly recalled in her memoirs starting out one morning over the Trail after all of the packing was done and the mules were hitched to the wagons: "The children were counted and loaded. A swift glance about to see that nothing was left behind and we were off for another day on the Trail. Drivers were calling, 'Get up there! Come along, boys!' Bull whips were cracking and all about the heavy wagons began groaning. The mules leaned into the collar and the great wheels began a steady creaking. Turn where we would, flocks of prairie chicken rose and went sailing across the open country."[15]

Others on the Santa Fe Trail wrote about seeing these birds during their journeys and about the large flocks. Among them was Josiah Gregg, who stated: "That species of American grouse, known west as the *prairie-hen,* is very abundant on the frontier, and is quite destructive, in autumn, to the prairie corn-fields."[16] The early settlers said there were millions and told of their fields being covered by them. Gregg also concluded from his own observations that prairie chicken and partridges, along with wild turkeys, geese, ducks, sandhill and white cranes, plover, curlew, hawks, and ravens comprised "most of the fowls of the Prairies."[17]

David Kellogg, an adventurer in the first wave of prospectors on their way to the Colorado goldfields, recorded in his diary on September 23, 1858, after reaching "the last outpost of civilization," Council Grove: "Prairie chickens were plentiful along the road yesterday."[18] Samuel Raymond, on his way to the Pikes Peak gold mines, made a brief note in his journal on April 19, 1859: "Boys went hunting for prairie hens and rabbits."[19] And Lewis Garrard noted seeing prairie chicken on a "detour through the prairie."[20]

In 1852, Dr. Michael Steck traveled to Santa Fe via the famous Trail to assume his new position as the superintendent of Indian Affairs for New Mexico Territory. He left from Independence, Missouri, on October 10 and reached Santa Fe on December 12, where he wrote a letter to a friend back home in Pennsylvania. Dr. Steck included a comment about prairie chicken: "We amused ourselves shooting grouse, which I can assure you are fine eating particularly when you superintend the cooking of them yourself."[21] Although considerable rain and snow were encountered, plenty of game birds were found along their route. In addition to prairie chicken, he mentioned ducks, geese, and turkeys, all of which they found "in great abundance."[22]

Dr. Steck was not the only Trail traveler to comment on "the fine eating" provided by prairie chicken. J. W. Chatham of South Carolina, for instance, made an entry in his private journal on June 6, 1849: "Soon after day[break] we had a light shower, wind S.W. warm and a fine breeze. I had the pleasure of breakfasting on Buffalo calf and Prairie chicken—both very fine and palatable. We all ate with wolf-like appetites."[23]

In 1851, Katie Bowen traveled the Trail from Leavenworth, Kansas, to newly established Fort Union in New Mexico with her husband, Captain Isaac Bowen. She wrote to her family and friends back home in Maine on June 28, 1851, at the Wakarusa River,[24] where their party had stopped because of heavy rains: "We are fixing up to spend the remainder of the day here and another night on this camp ground. The men servants managed to make us a cup of coffee in the rain, and as we had a quantity of cream & biscuit, we did not fare badly for breakfast. Now at noon the clouds are giving place gradually to the 'deep azure and gold' of a western sky and, oh sentiment, our Dinah [Katie's servant] has gone about making beans and roasting a fat prairie chicken that one of the teamsters just brought me. We do not fare so badly as some would imagine and you town people with fastidious appetites have no idea how well we relish our homely mess [army term for meal]."[25] On Christmas Day 1851, Katie and Captain Bowen hosted the first Christmas dinner given for the officers and wives at Fort Union. Meat dishes served included "a roast pig, saddle of venison, fillet of veal, and cold roast fowls with jellies."[26] Whether the cold roast fowls included prairie chicken is not known, although it would most likely have been included with other game fowl at such a fine feast, because prairie chicken were plentiful in that part of New Mexico.

James Francis Riley commented on prairie chicken in his recollections of freighting on the Santa Fe Trail and other trails west. These recollections were privately printed for family and friends by his grandson John Riley James in 1959. One hundred years earlier, in his descriptions of a westward trip in 1859, James Francis recalled: "Perhaps I had better tell you here (using the common bull whacker's language) what our grub consisted of. The first on the list was black coffee with sugar, next slapjacks or flat cake. Our only meat was called sow belly [salt pork] that was usually fried and the grease saved up to sop our bread in or make gravy or for shortening. These were the mainstays. . . . After we got out to where wild game was plentiful we fared much better. We could get plenty of prairie chicken and antelope. That year buffalo was scarce on our route. . . . Of course, we had most of the necessary seasonings to go with those things to make them palatable and best of all we had good strong appetites."[27]

Prairie chicken and other game fowl were roasted, stewed, broiled, or pan-fried by Trail travelers. The "necessary seasonings," referred to by James Francis Riley, carried among the provisions consisted of salt and pepper and a variety of spices, including cinnamon, cloves, ginger, nutmeg, allspice, and mace. According to Samuel P. Arnold, author of *Eating Up the Santa Fe Trail* (1990), in addition to spices sold in individual tin containers, seasoning mixtures were also available to wagon train cooks and travelers. One example was Dr. Kochpoder's special mixture, processed in Philadelphia, of salt, pepper, coriander, cloves, and other common spices.[28]

The word spread about the sweet, fine-flavored, medium-colored meat of the prairie chicken. It became a fashionable delicacy across the country and was served in fine restaurants in large cities in the United States and abroad, such as Chicago, New York City, and Paris. Market, or commercial, hunting was prevalent by the 1860s, lasting until the end of the 1800s. The numbers killed and sold are described as "beyond comprehension." Records of prairie chicken shipped to meat markets show that 513,000 were sold in Chicago in 1871, more than 600,000 were sold in New York City in 1873, and one large establishment in New York City sold 2,400 daily during the 1878 holiday season. More than 14,000 were shipped to Paris in 1875.[29]

––––––

The greatest numbers of prairie chicken occurred before the conversion of their native short- and mixed-grassland habitat to grazing range and cropland. The Dust Bowl during the 1930s caused a decline in their numbers, especially in southwestern Kansas, eastern New Mexico, and adjoining regions of other states where devastation was most severe. Their decline has also been caused by the eradication of native vegetation by spraying and removal, overgrazing, conversion to irrigated cropland, and the spread of towns and cities. Persistent drought, too, has taken its toll. Various sources estimate that the historical range for lesser prairie chicken has decreased by about 90 percent.

The Western Association of Fish and Wildlife Agencies and the Lesser Prairie Chicken Interstate Working Group have been developing a range-wide conservation plan that promotes voluntary conservation measures and land management practices in the range states of Colorado, Kansas, New Mexico, Oklahoma, and Texas. The effort involves numerous stakeholders: landowners, companies and industries (oil and gas, electric transmission, and windpower), state and federal game and range specialists, wildlife biologists, conservationists, bird-watchers, hunters, and others. The conservation plan's ultimate goals are to improve the present habitat and increase the amount of habitat available for these birds and, in doing so, increase their numbers.

The Lesser Prairie Chicken Interstate Working Group, wildlife agencies, and other entities were instrumental in establishing an annual aerial survey of the birds in 2012. The purpose of the surveys is to collect data on the bird's population across its range and on the location of leks, or breeding areas. The estimated rangewide population was reported as 18,747 in 2013 and 22,415 in 2014. Although a promising increase, the 2014 report noted that the numbers may fluctuate from year to year, mostly as a result of changes in grassland conditions with variations in rainfall. Also, the increase was not evenly spread over the four habitat regions existing in the five states. The sand sagebrush region where the Cimarron National Grassland is located showed a population decline. The surveys are invaluable in monitoring population trends and targeting areas where conservation measures are especially critical to restoring and expanding the habitat that this species requires for food and shelter.[30]

The Nature Conservancy has become involved in conservation of the lesser prairie chicken in the last decade with its purchase of 28,000 acres of prime grassland in eastern New Mexico and the establishment of the Milnesand Prairie Preserve, near the town of Milnesand, known for its lesser prairie chicken festivals. The preserve has more than fifty leks and excellent habitat for this species and many other grassland birds and mammals.

As early as 1995, the U.S. Fish and Wildlife Service was petitioned to list the prairie chicken as a threatened species under the Endangered Species Act. That did not happen until March 2014. Having a "threatened" listing means that the species may become in danger of extinction.

The future is uncertain for this species, as it is for others. But with strong support for and serious commitment to the objectives of the rangewide conservation plan, prairie chicken may very well continue their annual colorful courtship ritual on the grassland for years to come.

Rattlesnakes

Rattlesnakes are the subject of a huge body of lore, myth, and superstition formed and handed down over millennia. Profound fear of these snakes, and misconceptions about their behavior, commonly existed among Santa Fe Trail travelers. Many people following the routes of the Trail were encountering these reptiles for the first time. Misunderstanding and confusion about rattlesnakes are still common today.

Found only in the Western Hemisphere, rattlesnakes are pit vipers belonging to the large family Viperidae. They are not true vipers, which do not occur in the New World. The term "pit" comes from a sense organ visible externally as a facial pit below and back of the nostril. There are two genera, *Crotalus* and *Sistrurus,* in their subfamily *Crotalinae.* All species are venomous and have rattles. The genus *Crotalus* comprises the most species and the largest and most dangerous rattlesnakes, with ranges covering larger territory.[1]

The ranges of five species of rattlesnakes occur in the five states through which the Santa Fe Trail crossed: (1) the prairie rattlesnake (*Crotalus viridis viridis*)—central and western Kansas, Oklahoma Panhandle, eastern Colorado, and most of New Mexico; (2) the western diamondback rattlesnake (*Crotalus atrox*)—parts of northern and western Oklahoma and New Mexico; (3) the timber rattlesnake (*Crotalus horridus horridus*)—eastern Kansas and Missouri; (4) the desert massasauga (*Sistrurus catenatus edwardsi*)—southwestern Kansas, Oklahoma Panhandle, southeastern Colorado, and eastern New Mexico; and (5) the western massasauga (*Sistrurus catenatus tergeminus*)—southwestern, central and eastern Kansas, the Oklahoma Panhandle, and southeastern Colorado.[2]

The crucial characteristic distinguishing rattlesnakes from all other snakes, even from other pit vipers, is their rattle. Composed of loosely articulated, interlocking horny rings at the end of the tail, the rattle produces a distinct sound when vibrated. No other kind of snake has a rattle. Other snakes may have similar markings, be venomous, may coil and even vibrate their tails, but they are not rattlesnakes unless they have a rattle.

Such similarities among snakes caused confusion among Trail travelers, who would often kill any snake they came upon in their travels. J. Frank Dobie commented in his popular book *Rattlesnakes:* "The pioneer tradition towards the creatures of the earth was to kill them rather than to study them."[3] Today, people continue to kill snakes because they resemble rattlesnakes, although they are harmless and, like rattlesnakes, provide a valuable service for humankind by keeping disease-carrying rodent populations under control. One disease carried by rodents is hantavirus, which is caused by a virus spread by contact with rodent urine and feces, and which can be fatal.

Marc Simmons has written several articles about this reptile and remarked in his syndicated history column: "Rattlesnakes are not among my favorite creatures in this Southwestern homeland. But I'll be the first to admit that they add something of interest to the region."[4] In his cowpunching days, Simmons often heard their rattles when he drove cattle through thick vegetation, but he seldom saw them. "They would be on my mind, and sometimes when I had bedded down by the campfire I dreamed that a snake had crawled under my saddle, which I was using as a pillow. I'd wake suddenly and come flying out of the blankets. Then I would make a search to be sure that Mr. Serpent was not sharing my bed."[5]

Many Trail travelers, no doubt, awoke abruptly during the night or in the early morning in the same manner as Simmons, wondering if a rattlesnake had curled up in their blankets to keep warm. A few diaries and journals contain remarks about rattlesnakes being found much too close for comfort. Frank S. Edwards, a Missouri Mounted Volunteer, wrote in his book *A Campaign in New Mexico with Colonel Doniphan* (1847): "It was, by no means, an unusual occurrence for us, after a heavy dew, to kill in the morning within a quarter of a mile of camp more than twenty rattlesnakes, which, having come out to imbibe the dew, had become benumbed by the cool night air and, so, were an easy prey. Our Major awoke one morning with one of these reptiles coiled up against his leg, it having nestled there for warmth. He dared not stir until a servant came and removed the intruder."[6] Unfortunately, information was not provided by Edwards concerning how the servant was able to accomplish the task.

Drovers, mule-skinners, cowboys, and others on the Trail and throughout the Southwest used to encircle their blankets with a prickly hair rope laid on the ground, in the belief that snakes would not cross such a rope because the scratchy bristles tickled the snakes' bellies. Simmons remarked about this once popular practice: "I never put much faith in that old superstition. And besides, my up-to-date catch rope was made of smooth nylon, so it would not have been much use as a snake shield."[7]

Since the earliest days of the opening of the Trail, diaries, letters, reports, and other communications have commented on the presence of rattlesnakes. William Becknell, "the father of the Santa Fe Trail," noted the abundance of rattlesnakes during the inaugural journey that formally opened this great road of commerce. On September 1, 1821, Becknell and his small company of men started out from Franklin, Missouri, on their historic expedition. On September 24, he recorded that they had reached the Arkansas River and the world of the buffalo and the rattlesnake, "of which there are vast numbers here."[8]

On August 18, 1825, George Champlin Sibley, leader of the federal government survey of the Trail, wrote in his journal: "The immense plain in which we now are, and through which the Arkansas flows, is almost an uninterrupted level as far as the eye can [s]ee to the South and West. . . . There is not a single Tree anywhere to be seen on its banks. Our road this evening lay over some tolerably rough sandy ground, in which Rattle Snakes are very numerous."[9] He noted on the following day: "One of the Mules bit by a Rattle Snake this evening. These snakes are very numerous & troublesome here."[10]

From 1831 to 1840, Josiah Gregg saw countless rattlesnakes during his trips as a trader over the Trail. "Rattlesnakes are proverbially abundant upon all these prairies," he wrote, "and as there is seldom to be found either stick or stone with which to kill them, one hears almost a constant popping of rifles or pistols among the vanguard, to clear the route of these disagreeable occupants, lest they should bite our animals. As we were toiling up through the sandy hillocks which border the southern banks of the Arkansas, the day being exceedingly warm, we came upon a perfect den of these reptiles. I will not say 'thousands,' though this perhaps were nearer the truth—but hundreds at least were coiled or crawling in every direction. They were no sooner discovered than we were upon them with guns and pistols, determined to let none of them escape."[11]

Gregg related how the confusion caused by "the snake fracas" was worsened when a young wild mustang broke into their midst, and was further compounded by an ensuing skirmish between two mules in the area, one attempting to chastise the intruding colt and the other mule attempting to defend him. The reader can imagine the frenzy and noise deriving from such a scene, which Gregg described as "a capital scene of confusion." To restore order, the colt was shot and killed. The company pitched camp that evening opposite the "celebrated 'Caches,' a place where some of the earliest adventurers had been compelled to conceal their merchandise."[12] Doubtless there were lively conversations that evening, and for many days and evenings to come.

Gregg added that the rattlesnakes he encountered did not seem to be aggressive toward humans. "Scores of them are sometimes killed in the course of a day's travel; yet they seem remarkably harmless, for I have never witnessed an instance of a man's being bitten, though they have been known to crawl even into the beds of travelers. Mules are sometimes bitten by them, yet very rarely, though they must daily walk over considerable numbers."[13]

Susan Shelby Magoffin made very few comments in her diary concerning rattlesnakes, although she did refer to Gregg's experience in the sandhills along the Arkansas when writing at Big Coon Creek, where the caravan stopped at midday on July 13, 1846: "We also had a rattle-snake fracas. There were not hundreds killed tho,' as Mr. Gregg had to do to keep his animals from suffering, but some two or three were killed in the road by our carriage driver, and these were quite enough to make me sick."[14] She once confided that bugs, snakes, and mosquitoes were among the most disagreeable parts of her life on the prairie.

John W. Moore, pioneer and soldier, noted the number of rattlesnakes along the Trail, in particular along the course of the Arkansas River, which many others found to be where rattlesnakes were often seen. In his account, "The Santa Fe Trail Days," Moore described his first trip in 1867 at the age of twenty-one across the prairies and plains to Fort Lyon, located in southeastern Colorado: "We made a good finish to our trip. . . . And a lively finish it was, as on the divide between the Arkansas and the Smoky [Smoky Hill] rivers we ran into an immense lair of rattlesnakes that must have extended a mile. The reptiles were sunning themselves after a shower and lay at full length, sometimes crowded closely together, their sluggish bodies crossing each other and gleaming in the sun. We would shoot several before the rest would disperse, scuttling away over the warm sand with nothing but an ominous rattle and a sinuous motion on the green prairie to trace their course."[15] Moore made many trips along the Trail, delivering government supplies to forts. He became mayor of Kansas City in 1885 after the Trail had passed into history.

Alexander Majors, a prominent name connected to the freighting and mercantile firm of Russell, Majors & Waddell, formed in 1854, spoke at a meeting of "old plainsmen" in Independence, Missouri, in 1909. In regard to rattlesnakes biting animals, Majors stated: "The rattlesnakes on that road [Santa Fe Trail] in the beginning of the travel were a great annoyance, often biting the mules and oxen when they were grazing. At first, mules were used altogether for traveling, but they would either die or become useless from the bite of a rattlesnake, and the men would sometimes be sent ahead of the caravan with whips to frighten the snakes out of the pathway, but later on, the ox-teamsters, with their large whips, destroyed them so fast that they ceased to trouble [the oxen] to any great extent."[16] Simmons has provided additional detail: "Drovers walking beside ox teams kept an eye peeled for snakes and upon spotting one would skillfully take off the head with a well-aimed snap of their heavy bullwhips."[17]

Treatment administered to a mule bitten by a rattlesnake was included in the reminiscences left by James Brice, who arrived in Independence from Ireland in 1858 and was employed in the transporting of mail over the Trail. He recounted: "The herder brought one of the mules in off the pasture that was bitten by a rattlesnake. His head was twice its natural size. I brought a Cheyenne Indian to see the animal, who had me to throw [the mule] down and secure him from struggling. [The Indian] took my pen knife, sharpened the point of the small blade, tied a cord around it, leaving one-fourth inch uncovered, pricking the mule around the nostrils with the uncovered point, blood coming from each stab; rubbing the bloody part with gunpowder and telling me to keep him from water until the next morning. I complied with his instructions and the mule was all right next morning."[18]

———

Some hair-raising experiences along the Trail were recorded by men who had gone out on the prairie, sometimes alone, to hunt or to look after their animals. One was David Kellogg on his way to Colorado in the first wave of the gold rush there. He wrote in his diary on September 30, 1858, in camp at Cow Creek: "Today while crawling along a slight depression in the prairie to get inside the fringe of bulls which are always surrounding a herd of buffalo, I heard a sudden rattle just where I was about to put down my hand, and came face-to-face with a rattle-snake coiled for business. In my eagerness to stalk the buffalo I had not noticed him. I was thrilled as with an electric shock and, bounding to my feet, I placed my gun against the snake and blew him to pieces. It was an ungracious act on my part after he had given me fair warning, but I had but one thought in my mind and that was to kill that snake, and I was satisfied to see my cows [buffalo], one of which I had selected for my meat, go lumbering off over the plain."[19]

Lieutenant William B. Lane, stationed at Fort Union in 1857, encountered what he believed, at first, was a rattlesnake while hunting for antelope about fifteen miles from the fort. He caught sight of and approached a large herd, dismounted and picked up a stone to drive a picket pin (for hitching his horse) into the ground. Lane related his experience in a report written in 1894: "After about the third stroke with the stone I felt a sharp sting on the back of my left hand, and at the same moment heard the rattle of a snake, and saw within a few inches of my hand the last half of a large and horrid-looking rattlesnake just about to disappear in a hole in the ground. . . . I immediately examined my hand, and sure enough there were the two punctures, just the distance apart to correspond to the fangs of a snake. I was of course frightened almost out of my wits."[20]

Knowing he needed to get back to the fort as quickly as possible, Lane retrieved his picket pin from the ground, removed his brandy flask from his saddlebags and took a big swig. As he was heading back to the fort, Lane watched his left hand swell. He took another big swig, "a whopper" this time, and "raised both hands together that I might see just how much the left one was swollen. To my amazement the right one was just as large as the left, and not only that, there seemed to be several pairs of hands; in fact, the air was full of them, and all badly snakebitten. It suddenly occurred to me that I was *very drunk*."[21] The thought came to him that he could not have been badly poisoned, or the brandy would not have worked so quickly.

Lane struggled to stay upright in his saddle and began arguing with himself. He seemed to be two fellows—a sober lieutenant telling a drunken lieutenant that he "ought to be court-martialed for cowardice." Managing to stay in the saddle, Lane saw wild geese as he was drawing close to the fort, and thought he might "redeem himself" by shooting a goose to take back with him. Tying his horse, he proceeded toward the geese, when he noticed men approaching on horseback in the distance. He realized it would be better to head straight for the fort; his horse had come to the same conclusion and was trying desperately to break away. In mounting his pitching horse, Lane threw himself too far over his saddle and had to cling in that awkward position while his horse "went thundering across the plain . . . nearly a quarter of a mile" before Lane regained control.

Exhausted and relieved to be back at the fort, Lane concluded: "I had had what one might call 'a full day.' I had ridden over thirty miles, been bitten, as I supposed by a rattlesnake, got drunk and sober, was at the point of death and had recovered, and all this within twelve hours of the same day." He went to bed that night "feeling thankful that I had a back and a comfortable place to put it."[22]

Although Lane seemed to think that the snake that bit him was not a rattlesnake because he did not have the expected symptoms, he may well have been wrong about this. Lane's age, size, vigor, and good health could have enabled his body to resist the effects of the venom. Or he may have received a bite of weak venom, or even a dry bite, in which no venom was injected. Today, it is known that about one-fifth of rattlesnake bites are dry bites. Of approximately eight thousand bites from venomous snakes occurring each year in the United States, less than 1 percent result in death.[23]

———

Philip Gooch Ferguson of the First Regiment of the Missouri Mounted Volunteers recorded a snakebite cure similar to Lane's when another volunteer, David Russell, was bitten by a rattlesnake. Ferguson wrote in his diary on August 7, 1847: "Stopped at night a few miles past Middle [S]pring. Whilst hobbling his horse, Russell was bitten by a rattlesnake and became very much alarmed. One of the Mexicans cured him in the following manner. Made him swallow half a pint of whiskey, then tied a cord around his forefinger (the end of this finger being the place bitten) and cut it to the bone with a sharp knife, and then seared or burnt the wound. His hand was held down all night, and in the morning all danger was past. Some of [the] boys, seeing that whiskey was important in the cure of snake bite, complained to the Mexican that they were bitten and wished to be cured as he had cured Russell. The Mexican, discovering the ruse, wished to cut the flesh first and administer the whiskey afterwards!"[24]

Whiskey was the universal remedy for rattlesnake bites. There were many different folk and other remedies, including a variety of poultices made from plants, but the remedy of choice among the frontiersmen and others of that time was whiskey. Today, we know that alcohol causes bite victims more harm than good. It has the adverse effect of speeding up circulation, thereby increasing the rapidity of venom absorption. The large quantities of whiskey administered to snakebite victims during frontier days has led to the view that a number of deaths were caused by alcohol poisoning, rather than snake poisoning.

Lt. Lane's wife, Lydia Spencer Lane, wrote her recollections of the joys and sorrows of being the wife of an army officer on the frontier in her book, *I Married a Soldier; or, Old Days in the Old Army* (1893). Following their marriage in Pennsylvania in 1854, the couple headed for Jefferson Barracks, Missouri, and from there to a long list of posts stretching across Texas and the Southwest, with stops at several posts in New Mexico, including Fort Union. Among her recollections of her first journey to the western frontier, Lydia wrote: "I never went to bed without making a thorough search for a snake, tarantula, or centipede; but in all the years I spent traveling and camping, I never saw a snake about the tents . . . so that as time went on and I did not find the thing for which I watched, I grew careless, but not on that first expedition, when all was so new to me."[25] Lydia eventually adopted a live-and-let-live attitude: "Back of our quarters was quite a large yard, but there was not a living thing in it, except tarantulas, scorpions, centipedes, with the occasional rattlesnake for variety. As long as we left them undisturbed, they were harmless."[26]

A common misconception among travelers on the Trail was that rattlesnakes, prairie dogs, and small burrowing owls lived together in the prairie dog's burrow. As seen in chapter 3, many chroniclers made references to this widely held belief in their recorded observations of wildlife along the Trail. Matthew (Matt) C. Field, for example, wrote: "Snakes and owls are said to dwell with the dogs in their holes, with utmost harmony, but we had no opportunity of finding a proof of this, though the story is very generally believed."[27]

Alexander Majors elaborated on this misconception at the "old plainsmen's" gathering in 1909: "It has been claimed by men that snakes and prairie dogs, who were also found in great numbers upon the plains, lived in the same houses, the dog digging the hole and allowing the snake to inhabit it with him, but I do not think this is correct. Men came to this conclusion from seeing the snakes when frightened run into the dog-holes, but I think they did it to get out of the way of danger, and they lived, too, in the houses that had been abandoned by the dogs."[28]

———

A seemingly indestructible myth, held for centuries and first reported in print in the early 1600s, was known to people on the frontier. Sometimes referred to as "the snake-swallowing phenomenon," this myth has it that rattlesnakes swallow their young or take them into their mouths and release them when danger passes. Marion Sloan Russell, who had her own close encounters with rattlesnakes during her childhood, described such an occurrence involving her brother: "Once Will put his foot right down on a rattler. He was scared, but the snake did nothing. It did not even coil or strike out at him. This snake was certainly different. It lay still when Will removed his foot, lay still on the path to the milk house, and began slowly rattling its rattles. It fixed its great eyes on Will and flattened its head and stuck out its forked tongue and went on rattling its rattles. Suddenly from all around little snakes began to put in an appearance. There were a dozen or more of them. The old snake opened wide her jaws and the little ones disappeared down inside of her. When Will came to tell mother about what he had seen, she told him he had witnessed one of nature's miracles."[29]

Dobie collected numerous testimonies from people living in rural areas who claimed to have witnessed the same miracle. He trusted the veracity of some of the eyewitnesses, but declined to make a judgment regarding this subject, stating that his role was to transmit testimony and leave judgment to future observers. Biologists and herpetologists who have studied rattlesnakes for years have not reported witnessing this phenomenon and maintain there is no proof that rattlesnakes swallow their young.

An old fallacy prevalent among people today is that a rattlesnake's age can be found by counting the number of segments in its rattle. Bob Myers, Director of the American International Rattlesnake Museum in Albuquerque, New Mexico, said he finds this the most common misunderstanding among both adults and children who visit the museum. Rattlesnakes shed their skin from one to four times every year. With each shedding, a new segment appears. Over time, segments become brittle and break off. The only way a rattlesnake's age can be known is if someone happens to be present when a snake is born and a record is made at that time. Births of snakes in museums or zoos are recorded, but there is little chance of making such a record in the wild.

Rattlesnakes try to avoid contact with people. They bite as a defensive reaction, not as an act of aggression. They will readily defend themselves, in particular the western diamondback, if threatened. Their rattle is the only means they have to communicate a warning that danger is near and that people and other animals are treading on unsafe territory. Throughout human history, humans have been much more dangerous to them than they have been to people.

One can only guess to what extent attempts to eradicate rattlesnakes have depleted their numbers to the present time. It is known that their numbers are decreasing. This is unfortunate, because they play an important and integral part in the ecological web of life. Dobie stated: "Rattlesnakes, though not harmless, are the most interesting of all snakes in North America. I hate to think of the days when there won't be any."[30] He was pleased that more and more people, although still a minority, have learned to respect, even like, rattlesnakes as their understanding and knowledge of these fascinating creatures have grown. Dobie added: "I am good at belonging to minorities and I have joined this one. . . . Why should I pick on rattlesnakes?"[31] For that matter, why should anyone pick on rattlesnakes—or any other snakes—living along the old Santa Fe Trail and wherever else they range?

Chapter Eight

Grizzlies and Black Bears

Many people are surprised to learn that the grizzly once lived through-out the American West and on the Great Plains, including the grassy prairie, dry plains, river and valley bottoms, and mountainous region through which the routes of the Santa Fe Trail crossed. When the Trail opened to trade between Missouri and New Mexico in 1821, the grizzly's range extended from northern Alaska and western Canada to central Mexico. Adaptable to a wide variety of habitats, this member of the family Ursidae was seen, in particular, in riparian and mountainous areas thick with vegetation. Habitat changed with the seasons and the availability of food. The grizzly and the smaller black bear traveled from mountains to lowlands in spring, to higher elevations in summer, down to the lowlands in late summer and autumn, and returned to the mountains with the onset of winter.

Marc Simmons described the grizzly, one of the largest predators in the world, in his syndicated history column: "Of all the animals that once inhab-ited the great American West, none was regarded with more respect and fear than the grizzly bear. That surely was owing to his ferocious disposi-tion and huge size, the bears often weighing between five hundred and one thousand pounds. . . .

"Although grizzlies were abundant in mountainous regions where there was plenty of thick vegetation, they also were found in the lowlands. Wagon travelers heading toward the Southwest sometimes observed them on the plains. Late in the summer they leave the mountains for the open prairie, it is said by frontiersmen, searching for plants which they much relish."[1] Among those things relished were the plentiful wild berries, plums, and other fruit that ripened in the late summer along the creeks and streams. About 85 percent of the grizzly's diet comes from the native habitat's vegetation.

The name "grizzly" is used only for members of the brown bear family of North America. This species is divided into two subspecies: *Ursus arctos horribilis,* which translates as the horrible brown bear, and *Ursus arctos middendorffi,* which comprises the giant Kodiak bears of Alaska. "Grizzly" comes from the Old French word *grisel,* which means grayish. The grizzled appearance is caused by the silver-tipped guard hairs scattered throughout the bear's thick coat. Because of the silvery or whitish look of the coat, some early observers called them "white bears." In Mexico, this color was denoted in the name *el oso plateado,* or the silvery bear. Most grizzlies are medium brown, although colors range from a sandy blonde to a dark brown. Their greater size, a pronounced shoulder hump composed of muscle, and a large, wide, flattish face, sometimes described as "dish-shaped,"[2] distinguish them from other bears.

The black bear (*Ursus americanus*), or American bear, was also found in large numbers in the mountains and on the lowlands and was seen more frequently than the grizzly. Sometimes referred to as "the common bear," this bear's coat may also be brown, cinnamon, or lighter shades and is not as thick or long as the grizzly's. Smaller and less powerful, the black bear has larger ears and a more rounded muzzle and back. This bear adapts more easily than the grizzly to human activities.

In addition to being a trader, Josiah Gregg was an amateur naturalist. He included the black bear in his descriptions of animals on the Trail in *Commerce of the Prairies:* "About the thickety streams, as well as among the Cross Timbers, the black bear is very common, living chiefly upon acorns and other fruits. The grape vines and the branches of the scrubby oaks, and plum bushes, are in some places so torn and broken by the bear in pursuit of fruits that a stranger would conclude a violent hurricane had passed among them."[3]

Gregg also commented on the ferocity of the bears he had observed during his travels on the Trail from 1831 to 1840: "The black bear and grizzly bear, which are met with in the mountains, do not appear to possess the great degree of ferocity, however, for which the latter especially is so much famed farther north. It is true they sometimes descend from the mountains into the corn-fields, and wonderful stories are told of dreadful combats between them and the *labradores* [farmers]; but judging from a little adventure

I once witnessed, with an old female of the grizzly species, encountered by a party of us along the borders of the great prairies, I am not disposed to consider either their ferocity or their boldness very terrible."[4] Many men in the fur trade and others traveling the western wilderness held an opinion very different from Gregg's, however.

James O. [Ohio] Pattie, a youthful adventurer and trapper, also mentioned the damage caused by grizzlies in the fields of the far-flung settlements in his book *The Personal Narrative of James O. Pattie of Kentucky* (1831): "The country abounds with these fierce and terrible animals, to a degree, that in some districts they are truly formidable. They get into the corn fields. The owners hear the noise, which they make among the corn, and supposing it occasioned by cows and horses that have broken into the fields, they rise from their beds, and go to drive them out, when instead of finding retreating domestic animals, they are assailed by the grizzly bear. I have been acquainted with several fatal cases of that sort."[5]

Pattie, a traveler on the Trail and beyond from 1824 to 1830, told of a different fatal encounter with a grizzly that occurred when he camped in New Mexico. His servant accompanied another man outside to get a load of firewood, and a grizzly attacked the man, killing both him and his donkey. Those who witnessed this incident begged Pattie and his companions to help them kill the grizzly. He and three other men trailed the bear to its den and fired a gun. Pattie recounted the result: "The animal, confident in his fierceness, came out, and we instantly killed it."[6]

Other people taking the routes of the Trail southwest and west wrote about bears being killed near their campgrounds. John James Cleminson traveled with his large family from Independence, Missouri, to San Diego, California, after the Mexican War, when the number of emigrants on the Trail increased. He wrote in his diary on September 20, 1850, near Wagon Mound, New Mexico: "In the evening the hunters in advance [riding their horses ahead of the caravan] succeeded in killing a very large brown bear about two hundred yards to the right of our encampment. Also one of the hunters of the train in company [a wagon train traveling with Cleminson's] succeeded in killing another of smaller size. . . . We had several messes of bear meat, a new article of food to most of the company. Many of us ate it with good relish, while others could not eat it at all."[7] Cleminson's sparse descriptions may cause one to surmise that the first bear, "a very large brown bear," was a grizzly and the second bear killed was either a younger grizzly or a black bear. There is no way of knowing, which is sometimes the case with travelers' accounts. Those who spent more time in the region, such as fur traders and trappers, could easily distinguish which bear was seen or killed. It is not unreasonable that grizzlies could have been killed in this part of northern New Mexico in that year and season.

There are few accounts of a person being attacked by a black bear along or near the Trail, but Joseph Pratt Allyn told of one such incident in a letter written at Raton Pass in November 1863. Earlier that year, Allyn left his home in New York, traveled to Fort Leavenworth in Kansas and over the Trail via the Mountain Route to Santa Fe. His final destination was the Arizona Territory, where he assumed the position of associate justice for the Territory. After crossing the vast open plains, his party reached the Purgatoire River in southeastern Colorado. They stopped at one of the settlements strung out along the river, and the owner of a cabin on the road asked Allyn if there was a doctor in his party. His neighbor had been seriously wounded by "a cinnamon bear" three days earlier. Since there was no doctor at the settlement, someone had been sent to bring back medical assistance. Allyn explained the situation in his letter: "Two were out hunting, and both had shot and hit the bear, in the thicket, when out he bounded, eyes gleaming, and seized one and literally 'chawed' his leg up, while the other broke his gun all to pieces over the bear's head, and both were engaged in stabbing him until they killed him."[8] Allyn added in his letter that a doctor had finally arrived and took care of the injured man's wounds. The "cinnamon bear" usually refers to a black bear, but this bear may have been a grizzly.

———

The grizzly was first recorded for science by Captain Meriwether Lewis during the expedition of the Corps of Discovery from May 1804 to September 1806. Lewis and his co-captain William Clark recorded 122 animals and 178 plants unknown to science during their exploration of the Louisiana Purchase territory and search for a Northwest Passage. Astounded by the abundance and variety of wildlife, Lewis wrote in his journal that of all the fascinating animals recorded during their expedition, the grizzly was "by far, the most impressive of all creatures."[9]

The Corps of Discovery saw numerous black bears as they journeyed up the Missouri River, but not grizzlies. However, they often saw bear tracks that were too large to have been made by black bears. Although one member of the Corps had shot at a "white bear" [a grizzly] earlier and was forced to flee for his life, only a few had been seen, and all those sighted far off in the distance. Captain Lewis commented about this puzzling situation: "We have not as yet seen one of these animals, though their tracks are so abundant and recent. The men as well as ourselves are anxious to meet with some of these bears. The Indians give a very formidable account of the strength and ferocity of this animal, which they never dare to attack but in parties of six, eight or ten persons; and are even then frequently defeated with the loss of one or more of their party."[10]

Their wish to meet this elusive bear was fulfilled on April 29, 1805, when Captain Lewis and another member of the Corps killed, without much difficulty, a male bear that was not fully grown and estimated to weigh about three hundred pounds. Lewis noted in his journal: "It is a much more furious and formidable animal [than the black bear] and will frequently pursue the hunter when wounded. It is astonishing to see the wounds they will bear before they can be put to death. The Indians may well fear this animal, equipped as they generally are with their bows and arrows or indifferent fuzees [guns supplied by traders], but in the hands of skillful riflemen they are by no means as formidable or dangerous as they have been represented."[11]

Another grizzly was killed on May 5, and Captain Lewis's description of this bear and the difficulty experienced in killing the animal showed a marked change from his first impression. "Capt. Clark and [George] Drouillard killed the largest brown bear this evening which we have yet seen. It was a most tremendous looking animal, and extremely hard to kill notwithstanding he had five balls through his lungs and five others in various parts. He swam more than half the distance across the river to a sandbar and it was at least twenty minutes before he died. He did not attempt to attack, but fled and made the most tremendous roaring from the moment he was shot."[12] Lewis estimated that the grizzly weighed about six hundred pounds;

he measured eight feet, seven and one-half inches from nose to heel, five feet and ten and one-half inches around the chest, and three feet and eleven inches around the neck. The grizzly's foreclaws (Lewis called them "talons") were more than four inches long. Everything about this bear was beyond belief.

Lewis was clearly shaken by the experience. "These bears being so hard to die rather intimidates us all," he wrote in his journal. "I must confess that I do not like the gentlemen and had rather fight two Indians than one bear."[13] The grizzlies became such a persistent threat that the Corps was ordered to travel in pairs and keep their guns close at hand. This was a sound order, because over the following months they were chased back and forth over the prairie and were forced to climb trees, plow through thorny thickets, and jump into the river to escape the bears' ferocity. There were some very close encounters, but no one on the expedition was seriously injured by a grizzly. In spite of all their difficulties, the Corps killed forty-three grizzlies during the expedition. No mention was found of the total number of black bears killed.

In 1805, Lewis and Clark sent several live animals captured on the expedition to President Jefferson. Two years later, Captain Zebulon Montgomery Pike sent Jefferson "a variety of curiosities" collected during his own expedition. Among the curiosities he sent were two grizzly cubs, male and female, which were shipped down the Mississippi to New Orleans, then to Baltimore, and finally to Washington, D.C.

Pike explained in a letter to President Jefferson that the cubs had been purchased from an Indian soon after their birth near the Continental Divide. They were so small that Pike's men on horseback carried them in their laps. After a few days, the cubs were put in a small cage and carried between two packs on the back of a mule. The men played with them, treating the little grizzlies like pets. Pike assured the president that they were "perfectly gentle knowing no other benefactor than man from the time of their birth."[14]

In a letter written on November 6, 1807, President Jefferson informed Pike that the cubs had arrived in good health, but he did not express the delight he had when a prairie dog and a magpie were received alive from Lewis and Clark. The president told Pike that the cubs would be sent to Charles Willson

Peale, artist and curator, in Philadelphia, the same man who had received the numerous specimens sent from the West in 1805. In February 1808, Peale put the grizzlies on public exhibit for a brief period; they were almost a year old at that time, and each weighed about seventy pounds.

As the little grizzlies grew large and fearsome, they attacked anyone who came within their reach. It was reported that "a teasing monkey had an arm and shoulder blade torn off by a sweep of the great claws."[15] When one of the bears broke loose from the cage and stalked the basement of the building where they were kept, Peale secured the basement door and shot the errant bear the following morning. He also killed the other grizzly still in the cage. Both were mounted for exhibit.[16]

———

A myth passed down from those early days on the frontier wilderness and echoed by travelers on the Santa Fe Trail was that one in danger from a grizzly should climb a tree (if available) to save one's life. The belief was commonly held that grizzlies could not climb a tree because their long, slightly curved foreclaws made it impossible for them to get hold of the trunk and because their size and weight would keep them grounded. While some people were able to save their lives by climbing a tree, others were not so lucky. In fact, an angry grizzly can shimmy up at least the first ten feet of a tree trunk. His bulk does not allow him to go much higher or to remain off the ground for long, but with his long, powerful reach, the grizzly can grab a climber who isn't high enough in the tree. When a photographer was treed by an enraged grizzly in Glacier National Park in 1987, the grizzly followed him eighteen feet up the tree, dragged him down, and killed him.[17] The smaller, more agile black bear is a very good tree climber, able to climb much higher, and will readily climb a tree for resting and napping or to have a safe haven when afraid, chased, or threatened.

A frontiersman who managed to save his life by climbing a tree was Richens Lacy "Uncle Dick" Wootton, who built and operated a toll road for wagons traveling on the Mountain Route and going over the Raton Pass, which connects southeastern Colorado and northeastern New Mexico. Wootton saw many grizzlies and black bears during his years in the Raton Mountains, where black bears still live to the present day. In one encounter, he had to take refuge in a pine tree after shooting at a mother grizzly and missing. Uncle Dick, however, did not think the grizzly was a man-killer by nature, declaring in defense of the bear: "That he has been guilty now and then of staining his 'chops' with human gore is true, but it was usually under circumstances which would have made 'justifiable homicide' a proper verdict, if the affair had been between man and man."[18] In those days on the frontier wilderness, it was common knowledge that a mother bear with cubs was very dangerous and more likely than not to chase away or attack hunters and others who came too close. But most men on the frontier and in the mountains would shoot at any bear they saw or came upon within range of their guns.

Bears are excellent swimmers, and grizzlies, in particular, can swim a long distance. Thus, jumping in the river does not mean that a person has found certain safety. Although once thought to have poor eyesight, another grizzly myth, their eyesight is similar to a human's, and their hearing is slightly better. Able to pick up sounds of a human voice from a quarter-mile away, it is understandable that the grizzly is often long gone before humans arrive on the scene. This explains why their fresh tracks, but not the bears themselves, were often seen by explorers and others.

The grizzly's most important sense is his keen sense of smell. That particular natural gift almost spelled the end for Captain Ezekiel Williams, a trapper whose adventures in the Rocky Mountains, after Pike's 1806–1807 expedition, made him an authority on that part of the West. He also had important connections to the Santa Fe Trail.

Captain Williams, as the story goes, led a party of twenty trappers that set out from St. Louis in the spring of 1807 or 1809 for the upper reaches of the Missouri River. Only Williams and two of the other men were still alive when the party reached the upper region of the Arkansas River some months later. Not sure of their location, the three decided to separate. Williams chose to head downstream in his roughly made canoe, trapping beaver wherever the stream took him. Traveling mostly at night, he continued downstream on the Arkansas through country unknown to him. He did not use his gun unless it was absolutely necessary to procure meat. On one such occasion, having been unable to trap a beaver for his dinner, he shot a deer, and after eating his fill he placed the remainder of the carcass in the canoe. He usually slept in the canoe, which he moored to the shore with a piece of rawhide about twenty feet long. This allowed him to make a quick getaway by cutting the string and gliding away without making a sound.

One night, the captain was awakened by a loud tramping noise. He was able to see a huge grizzly come down to the water and start to swim toward the canoe. The grizzly had smelled the deer carcass. Grabbing his ax, Williams stood with it uplifted, ready to drive his weapon into the grizzly's head. When the grizzly reached the canoe, he put both forepaws upon the end where the carcass was stashed, nearly overturning it. The captain struck a paw with the ax, which caused the bear to withdraw that paw, but he held onto the canoe with the other. With all his strength, Williams struck the bear in the head, causing the animal to sink into the murky water, never to be seen again. The following morning, he found in his canoe two bear claws that had been cut off by the ax. He kept the claws as trophies to show to others whenever he recounted that experience with the grizzly.

The rest of the story tells about Williams being taken captive by Kansas Indians, who confiscated his furs and other belongings. During his captivity, he learned that the Indians planned to go to Fort Osage on the Missouri River. He was eventually given his liberty to proceed on his trip downriver, but the furs were not returned. Realizing the Indians would sell the furs when they reached the fort, he headed for the Missouri and arrived in the area near the fort before the Indians. The government trader, or factor, at Fort Osage at that time was George Champlin Sibley, who would later lead the commission appointed in 1825 to survey the Santa Fe Trail. Sibley refused to pay the Indians their annuities until the stolen goods were returned to Williams. The persevering trapper headed for St. Louis, where he sold the furs. It was 1813 when he arrived in St. Louis and that long, arduous adventure was finally ended.[19]

Englishman and intrepid adventurer George Frederick Augustus Ruxton traveled far and wide through New Mexico, the Greater Southwest, and Mexico. In his book *Adventures in Mexico and the Rocky Mountains,* Ruxton recorded his impressions of the grizzly: "His great strength and wonderful tenacity of life render an encounter with him anything but desirable. . . . Although, like every other wild animal, he usually flees from man, yet at certain seasons, when maddened by love or hunger, he not unfrequently charges at first sight of a foe; when, unless killed dead, a hug at close quarters is anything but a pleasant embrace, his strong hooked claws stripping the flesh from the bones as easily as a cook peels an onion. Many are the tales of bloody encounters with these animals which the trappers delight to recount to the 'greenhorns,' to enforce their caution as to the fool-hardiness of ever attacking the grizzly bear."[20]

Some of these tales also provided cautionary lessons about the paramount importance of keeping a good rifle at hand when in grizzly country. Colonel Henry Inman illustrated the point in his story about a careless trapper somewhere along the Arkansas River in the early 1840s. "Several years before the acquisition of New Mexico by the United States, two old trappers were far up on the Arkansas near the Trail, in the foot-hills hunting buffalo, and they, as is generally the case, became separated. In an hour or two one of them killed a fat young cow, and, leaving his rifle on the ground, went up and commenced to skin [the buffalo]. While busily engaged in his work, he suddenly heard right behind him a suppressed snort, and looking around he saw to his dismay a monstrous grizzly ambling along in that animal's characteristic gait, within a few feet of him.

"In front, only a few rods away [one rod equals 16.5 feet], there happened to be a clump of scrubby pines, and he incontinently made a break for them, climbing into the tallest in less time than it takes to tell of it. The bear deliberately ate a hearty meal off the juicy hams of the cow, so providentially fallen in his way, and when he had satiated himself, instead of going away, he quietly stretched himself along side the half-devoured carcass, and went to sleep, keeping one eye open, however, on the movements of the unlucky hunter whom he had corralled in the tree. In the early evening his partner came to the spot and killed the impudent bear, that, being full of tender buffalo meat, was sluggish and unwary, and thus became an easy victim to the unerring rifle; when the unwilling prisoner came down from his perch in the pine, feeling sheepish enough. The last time I saw him he told me he still had the bear's hide, which he religiously preserved as a memento of his foolishness in separating himself from his rifle, a thing he has never been guilty of before or since."[21]

A sobering account of an encounter with an enraged grizzly was told by James Ohio Pattie in his published narrative. Pattie is known to have had a tendency to exaggerate, embellish, and sometimes even fabricate his stories, so the following may be a reliable account or a cautionary tale made up by him. He wrote about one of the men in his party being severely injured by a grizzly: "The growls of the bear, as he tore up the ground around him with his claws, attracted all in his direction. Some of the men came so near, that the animal saw them, and made towards them. They all fired at him, but did not touch him. All now fled from the furious animal, as he seemed intent on destroying them. In this general flight, one of the men was caught. As he screamed out in his agony, I, happening to have reloaded my gun, ran up to relieve him. Reaching the spot in an instant, I placed the muzzle of my gun against the bear, and discharging it, killed him.

"Our companion was literally torn in pieces. The flesh on his hip was torn off, leaving the sinews bare, by the teeth of the bear. His side was so wounded in three places, that his breath came through the openings; his head was dreadfully bruised, and his jaw broken. His breath came out from both sides of his windpipe, the animal in his fury having placed his teeth and claws in every part of his body. No one could have supposed that there was the slightest possibility of his recovery, through any human means. We remained in our encampment three days, attending upon him, without seeing any change for the worse or better in his situation. He had desired us from the first to leave him, as he considered his case as hopeless as ourselves did. We then concluded to move from our encampment, leaving two men with him, to each of whom we gave one dollar a day, for remaining to take care of him, until he should die, and to bury him decently. . . .

"We set off, taking, as we believed, a final leave of our poor companion. Our feelings may be imagined, as we left this suffering man to die in this savage region, unfriended and unpitied. We traveled but a few miles before we came to a fine stream and some timber. Concluding that this would be a better place for our unfortunate companion, than the one where he was, we encamped with the intention of sending back for him. We dispatched men for him, and began to prepare a shelter for him, should he arrive. . . . We set traps, and caught eight beavers, during the night. Our companions with the wounded man on a litter, reached us about eight o'clock at night.

"In the morning we had our painful task of leave taking to go through again. We promised to wait for the two we left behind at the Arkansas river. We traveled all day up this stream. I counted, in the course of the day, two hundred and twenty white [grizzly] bears. We killed eight that made an attack upon us; the claws of which I saved. Leaving the stream in the evening we encamped on the plain. A guard of twenty was relieved through the night, to prevent the bears from coming in upon us. Two tried to do it and were killed."[22]

The severely mauled trapper succumbed to his injuries a few days later and was buried. Pattie's company had camped in far western Kansas near the Colorado boundary. His count of 220 grizzlies seen in a single day has been frequently questioned and considered by some historians and others as an exaggeration or error on the part of either Pattie or his publisher. But even Pattie's skeptics consider the count as likely evidence that a large number of grizzlies once lived and roamed in the western country.

The harrowing saga of trapper Hugh Glass has been told by George Frederick Augustus Ruxton and others up to the present time.[23] Glass joined a large trading venture that ascended the Missouri River. He had a terrible scrape with a grizzly in the autumn of 1823 near the forks of the Grand River in present South Dakota. Unlike Pattie's companions, who gave succor to their badly injured friend, Glass was left to die alone by fellow trappers. Although two had promised to stay behind with the injured trapper and bury him when he died, they departed with Glass's possessions, including his guns and moccasins. After catching up with the other trappers, the two men told them that Glass had died and was buried. Glass, however, regained consciousness and managed to crawl for many miles, surviving on berries and roots. He was later able to fashion a crude raft for himself and eventually reached Fort Kiowa, two hundred miles from where he had been mauled by the grizzly. He tracked down the two men who had left him, but he did not avenge the grievous wrong done to him. One can imagine the shock and humiliation of those men. In later years, Hugh Glass was a hunter at the Fort Union Trading Post of the American Fur Company (not connected to Fort Union on the Trail) in present North Dakota. He was killed by Indians in 1833.

———

The last decade of the Trail, from 1870 to 1880, saw a tremendous influx of people and goods westward. Towns, ranches, and farms spread across the land, and the livestock industry expanded significantly. Wild animals, great and small, were under siege. They were being hunted and killed in large numbers, their natural habitats were being destroyed, and their food sources were disappearing. Grizzlies and black bears retreated to timbered mountains and more remote wilderness areas. They were a frightful nuisance to the newcomers, who demanded the eradication of large predators and other wild animals, all looked upon as detriments to economic and social progress.

Simmons described the situation as it affected the grizzly: "Since this bear was regarded by man with about the same affection he held for the rattlesnake, it is not surprising that all-out war was waged upon the lordly grizzly. There was no one, it seemed, who was willing to come to his defense. . . . Cattlemen, sheep raisers, and homesteaders all were loud in their call for extermination of the grizzly. Wolves and prairie dogs also were placed on the death list. Congress responded and in 1914 set up a special agency under the Department of Agriculture to eradicate these 'varmints.'"[24] Government, bounty, market, and other hunters shot, trapped, and poisoned grizzlies and other wild animals in astounding numbers. In the late 1920s, it was estimated that only twenty-eight grizzlies were left in the national parks of the Southwest; in 1936, there were only ten. The 1940s witnessed the extermination of the grizzly in New Mexico and Arizona. The last grizzly in southern Colorado was killed in 1979.[25]

Black bears also declined significantly. In 1924, the U.S. Forest Service estimated that only fifteen hundred were living in the national forests of New Mexico, southern Colorado, Arizona, and southern Utah. These adaptable, opportunistic omnivores, however, made a comeback. They moved into areas vacated by the grizzlies and to locations providing more shelter and wild foods for them and their young. Today they are seen in towns and other populated areas, especially during the summer season in years when precipitation has not arrived early enough to spur the growth of plants on which they depend. About 70 percent of their diet is composed of the harvest bounty from plants and trees.

In the 1890s and early decades of the twentieth century, as the old Santa Fe Trail slipped into oblivion many of the ruts and swales became covered with low vegetation or were turned into farming and grazing land. Like the Trail, memories of grizzlies and black bears roaming the vast, open, untamed plains faded away and were forgotten. Today, the black bear is New Mexico's official state mammal.

The most severe problem facing the grizzlies, the black bears, and other wildlife is habitat loss. In his epilogue to *The Valley of the Grizzlies* (1998), nature writer and photographer Robert H. Busch states: "All life has bounds, constraints that put some type of restrictions on our lives. But for wild animals, man-made constraints now rule their wild worlds. Grizzlies, like most other animals, are being pushed into smaller and smaller habitats. Soon they may have nowhere at all to go."[26]

Mustangs—The Wild Horses

In the early 1830s, the Santa Fe Trail was young, the land was pristine, and magnificent droves of wild mustangs roamed the grassy plains in overwhelming numbers.

The mustang, or wild horse (*Equus caballus*), member of the Equidae family, was one of the preeminent animals seen on the prairies by the travelers following the routes of the Trail. And indeed, mustangs were the topic that Josiah Gregg chose to write about first in his treatise on animals in *Commerce of the Prairies*. He explained at the beginning of his essay why he chose the mustang: "By far the most noble of these [animals], and therefore the best entitled to precedence in the brief notice I am able to present of the animals of those regions, is the mustang or wild horse of the Prairies."[1] Gregg then provides a vivid image, subsequently quoted in many sources, that gives a rough estimate of their numbers and an insight into what seeing the mustangs might have meant to Trail travelers: "Large droves are very frequently seen upon the Prairies, sometimes of hundreds together, gamboling and curvetting within a short distance of the caravans."[2]

In the autumn of 1831, the same year Gregg set out on his first journey over the Trail, Albert Pike, a young, well-read Bostonian and distant relative of explorer Zebulon Montgomery Pike, left from Independence, Missouri, in a caravan headed for Taos and captained by Charles Bent, trader and partner in Bent's Fort. Albert's journals, filled with poetry and prose, were published in 1834, placing them among the first Trail reminiscences published as a book and predating Gregg's book by a decade. Albert Pike was deeply moved by the mustangs and their beauty. He wrote in *Prose Sketches and Poems, Written in the Western Country:* "Hardly a day passed without our seeing a herd of them, either quietly feeding, or careering off wildly in the distance. They are the most beautiful sight to be met with in the prairie. Of

all colors, but more commonly a bay [reddish brown], and with their manes floating in the wind, they present a beautiful contrast to the heavy, unwieldy herds of buffalo, which seem, even at their best speed, to be moved by some kind of clumsy machinery."[3]

J. Frank Dobie, writing a century later, described the mustang as "the most picturesque wild species of the land."[4] He was born in 1888 in Texas, where he "began to learn to ride before he finished learning to walk," even though his birth "came after the day of the mustang had passed."[5] Dobie observed wild horses in various locales where some still lived. For several decades, he collected hundreds of accounts of those bygone days from old settlers, cowboys, and mustangers. He probably wrote more about wild horses than any other writer past or present.

Mustangs had a variety of names: buffalo horse, Indian pony or Indian horse, Spanish pony, cow pony, range horse, saddle horse, and "Cayuse," a name borrowed from the Cayuse Indians, who lived in parts of what are now eastern Oregon and Washington. They had many horses and traded them with other tribes and sold them to travelers on the Oregon Trail. Their name became popularly used in the West for cow ponies or range horses used by cowboys. The name "mustang" is an English corruption of the Spanish word *mesteño* (feminine *mesteña*), a word with a long history in Spain before it came into use in the New World. *Mesteñeros,* or mustangers, were the men who hunted, captured, and killed mustangs for profit.

This hardy animal was tough and well-built with excellent feet (hard hooves) and sturdy legs. Mustangs were generally of smaller size that stable-bred horses because the hardships of their wild lives and the severe weather on the plains stunted and changed them over time. Their height ranged from thirteen to sixteen hands, averaging fourteen hands. "Hand" is a measure equivalent to four inches (14 hands = 4 feet 8 inches). Technically, any horse up to 14.2 hands, or 4 feet 10 inches at the shoulder, is considered a pony; one measuring over 14.2 hands is considered a horse. Their weight was generally from seven hundred to one thousand pounds.[6] The basic social unit, sometimes called a harem band, consists of a dominant stallion, several mares, and juveniles.

All the pretty wild horses on the vast grassy plains made an eye-catching, unforgettable sight. In addition to considerable variety in conformation, the mustangs came in all colors of the equine spectrum: bay, black, brown, buckskin, chestnut, dun, gray, grullo, palomino, roan, sorrel, and white.[7] Many had coats of solid color or a shade of a specific color, while others had beautiful markings and patterns of one color or shade on white coats, such as the paints and pintos. The Indians of the Plains prized all of their horses for their hardiness, loyalty, and intelligence, and particularly valued the paints for their colors.

The mustangs' appearance, however, was considered by some to be less than appealing. Gregg once remarked: "Their elegance has been much exaggerated by travelers, because they have seen them at large, abandoned to their wild and natural gaiety. Then, it is true, they appear superb indeed; but when caught and tamed, they generally dwindle down to ordinary ponies."[8] In spite of their size and rugged appearance, Dobie insisted the mustang had as much savvy as any horse that ever lived. Stanley Vestal, author of *The Old Santa Fe Trail* and a contemporary of Dobie, added: "Once broken, the mustang showed as much sense as a mule, as much endurance as the wolves that pestered him, and he had a 'hard' stomach that enabled him to stand an incredible amount of riding, and to go without grass and water for long periods. He could dodge an angry buffalo—or a man's snaring rope; he would buck and kick and bite. But he could go, and go, and go."[9]

Joseph Pratt Allyn discovered this to be the case when he purchased a mustang in 1863. He traveled the Trail from Kansas to New Mexico via the Mountain Route and continued his journey to Arizona. His long, informative letters written along the way were published in the *Hartford, Connecticut Evening Press*. In one written in December 1863 at Fort Wingate in northwestern New Mexico, Allyn confided that he shed a tear when he parted with his faithful, stable-bred horse, worn out and in need of rest and good food after the long ride. Before continuing to Arizona, Allyn acquired two horses, one of these being a mustang described as "a grey pony, native to the country, that keeps fat on grass, scarcely knows what corn is, a wild, frisky, treacherous little scamp, hard to saddle, stubbornly opposed to being rode and to going fast (unless it suits his freak), and then he can run."[10]

Vestal stressed the importance of the horse to Trail travelers: "A man's first thought on reaching a prairie port was to obtain an animal to carry him to Santa Fe. . . . As with arms and clothing, the greenhorn found that he had everything to learn when it came to buying a saddle animal. Horses bred in the States did not stand up well under prairie conditions; horses native to the Plains were a hardy breed."[11] He added: "As likely as not, his name was Paint. The greenhorn who bought him, merely because he looked western and bizarre, soon found that he had an animal with any number of good points—and a mind of his own."[12]

Lewis Garrard, a self-confessed greenhorn from Ohio, arrived in Westport, Missouri, in the summer of 1846. He waited there for the arrival of Céran St. Vrain, leader of the company in which Garrard traveled the Trail. While waiting, Garrard and a new acquaintance rode on hired horses from their encampment into bustling Westport to look at horses. Among the fascinating sights he saw were "different Indians, in fanciful dresses, riding in to trade and look around on their handsome ponies."[13] When St. Vrain arrived on September 1, Garrard finally had to purchase a horse of his own for the long journey. The wagonmaster of the caravan, Frank DeLisle, called *le maître de wagon* by Garrard, sold him a horse for $50. He described his new steed as having "a fanciful color, brown and white spots, and white eyes" and, as one might guess, named him "Paint." Garrard thought it was "a descriptive though not euphonious name."[14]

According to Garrard, Paint was "a noted buffalo chaser," and he "anticipated much excitement through [Paint's] new services." Whether or not Paint was really a bona fide buffalo chaser, he admitted: "There was plenty of excitement and considerable frustration. I have worked myself into a profuse perspiration, with vexation, a hundred and one times in vain attempts to trap him."[15] On those occasions, Garrard ended up hanging his saddle on a wagon and walked with others along the Trail.

It was apparent that learning how to use a *lazo* (Spanish for lasso) was a necessity on the frontier, and especially around mustangs. Luckily, there was assistance available to the greenhorns. Garrard remarked: "The maneuvers of the Mexicans of our company are really astonishing in lassoing unruly mules and horses; dodge as they may, or run about, the lariat noose is sure to fall on the unwilling necks; a loop thrown over the nose, the gagging Spanish bit forced into the mouth, the saddle clapped on, and the rider firmly in that, with galling spurs tickles the side ribs, and flies and curvettes on the plain in less time than it can be written."[16] Garrard and Paint finally grew accustomed to each other and were together until January 1847. Freezing weather and lack of forage had turned Paint into "a poor, old and broken down Pinto," forcing Garrard to trade him for a "a raw-boned, impetuous piece of mule flesh" with the ominous name of Diabolique.

Dobie emphasized in his telling of the mustangs' story that the prairies were natural horse country, and mustangs were prairie animals by nature. The land was perfectly suited to their needs, providing a dry climate, native grasses on which they thrived, and plenty of space to run freely and swiftly outdistance predators. Such excellent suitability caused the mustang population to multiply very rapidly. According to Dobie, the wild horse population probably reached its zenith at the end of the Mexican War in 1848 and numbered at least two million. "Even after tens of thousands had been captured or killed, many remained until the land was all fenced in."[17] Vestal also commented about the plains being prime habitat for mustangs: "The great weakness of the horse lies in his feet, and the chief enemy of his feet is moisture. Keep a horse dry-shod, with plenty of forage, and the world is his paradise. Such a paradise was furnished, ready-made for him on the high, grassy, semi-arid Plains of the West."[18]

––––––

The mustangs' history on the North American continent has been researched and debated for many years. Most animal scientists agree that North America was the ancestral home of the genus *Equus,* beginning with its earlier form, the Dawn Horse, or *Eohippus* (now called *Hyracotherium*). Based on the fossil record going back about sixty million years, paleontologists have named at least sixty species of the genus *Equus* from North America.[19] Horses roamed North America for millions of years until about ten or eleven thousand years ago when they and other larger mammals, except the pronghorn and a few others, became extinct.[20] The horses that had ventured earlier across the Bering land bridge and spread westward across Asia to Europe and North Africa did not become extinct.

The Spaniards brought *Equus* back "home" to his roots in North America. The modern horses that accompanied the *conquistadores,* however, were domesticated and different from any North American ancestor. They had evolved over the centuries, and horses in Spain had undergone selective breeding with the Barbs of the Moors, producing the famous Andalusians that came with the Spaniards. Horses arrived on the ship carrying Hernán Cortés and his soldiers in 1519, when they arrived in the Yucatán with the goal of acquiring new lands, slaves, and riches. Cortés was the first to bring cavalry to the North American continent, arriving in Mexico with sixteen stallions and mares, described as fourteen Arabians and two Andalusians. Also, one foal was born abroad ship. Because of the meticulous records kept by Spanish chronicler Bernal Díaz del Castillo we have descriptions of those horses, including their colors (bay, chestnut, brown, gray, black with spotted or dappled pattern, and sorrel).[21]

Dobie called the horses of the *conquistadores* "the guarded atom bomb of the Spaniards."[22] Shock and awe gripped the Indians when they first saw the horses. Cortés and the men who arrived after him understood the power of their horses to intimidate and subdue. They encouraged the Indians to believe that their horses were "monsters that devoured human flesh," and Indians were not allowed near the horses. After the conquest of Mexico, one of the first ordinances passed prohibited any Indian from riding a horse. This prohibition was in effect long after Mexican *vaqueros* (cow-workers), the first cowboys, were riding horses.

When the expedition of Spanish explorer Francisco Vásquez de Coronado arrived in New Mexico in 1540, the Indians called the Spaniard's horses "big dogs." The horses carried the Spaniards' gear and equipment, similar to the much smaller dogs that carried or hauled the Indians' domestic and hunting gear on their backs or on travois, the carrying platform dragged along the ground behind them. As time passed and the Indians of the Plains were able to obtain horses, their lives were profoundly transformed. Horses transported them on the move or on the hunt, whereas before they had always traveled and hunted on foot. Their mode of life changed dramatically, "from crop-bound camps to a boundlessness limited only by the winds of winter, the drift of buffalo, the fruiting of berries, fresh grass (which was nearly everywhere in season), and cottonwood when the grass was ice-locked."[23] Horses made it possible for them to reach impressive heights in the arts of horsemanship, hunting, and warfare.

It was long thought that mustangs living on the plains when the European Americans arrived in the 1800s had originated from horses brought to the Southwest by Coronado. Marc Simmons has written about this long-held belief: "For years popular belief held that the huge herds of horses that ran free on the Great Plains were descended from animals that escaped from Coronado's Expedition in 1540. But since the Spaniards reported no losses and wild horses were not seen until the 1700s, that theory no longer holds water. We now think that horses escaping from ranches in northern Mexico in the middle colonial period found their way to the pastureland of the Llano Estacado [Staked Plains] in eastern New Mexico and western Texas, where they rapidly multiplied."[24]

In addition to the horse-breeding ranches of northern Mexico, there were other "seminal seedbeds of horse diffusion" in the Southwest. An important one was the abandonment of Spanish horses following the Pueblo Revolt of 1680, when the Spaniards were forced out of New Mexico and did not return until 1692. During that period, the Indians of New Mexico began capturing horses and trading them to Plains Indians and other tribes of the Southwest. Another "seedbed" was the abandonment and escape of horses from the short-lived missions in East Texas in the 1690s.[25]

Over a period of time, those horses from northern Mexico, New Mexico, and Texas worked their way northward throughout the Great Plains. They were joined by a multitude of other horses that ran away from or became separated from exploring parties, from the *caballadas* (loose herds of horses, mules, and cattle) of the caravan companies on the Santa Fe Trail and other trails, from cavalry herds, and from encampments and settlements. Horses, as well as other livestock, also fled in fright from storms and lightning, raiding Indians, prairie fires, and wolves and other predators.

Stampeding mustang and buffalo herds also carried off many animals. Gregg observed: "It is sometimes difficult to keep them [the mustangs] from dashing among the loose stock of the traveler, which would be exceedingly dangerous, for once together they are hard to separate again, particularly if the number of mustangs is much the greatest. It is a singular fact, that the gentlest wagon-horse, . . . once among a drove of mustangs, will often acquire in a few hours all the intractable wildness of his untamed companions."[26] Thus, feral horses (once domestic, but returned to wildness) added to the growing mustang herds. Trail diaries and journals are filled with accounts of animals running away or being driven off and of the efforts expended by people in the caravans to retrieve them, often discouragingly unsuccessful.

When European Americans arrived at the beginning of the 1800s, wild horses were present in great numbers across the Plains. During his 1806–1807 expedition, Captain Zebulon Montgomery Pike wrote about seeing them numerous times on his way up the Arkansas River and in Texas. The exploring party camped on October 28, 1806, at the mouth of the Pawnee River [Fork] where it entered the Arkansas River from the north at Larned in present-day Kansas. The following day, Pike recorded his first sighting: "About noon discovered two horses feeding with a herd of buffalo; we attempted to surround them, but they soon cleared our fleetest coursers. One appeared to be an elegant horse; these were the first wild horses we had seen."[27]

On November 1, Pike wrote in his journal about seeing a herd of horses on the prairie as he was viewing the land with his spyglass. He and others rode out to view them: "When within a quarter of a mile, they discovered us, and came immediately up near us, making the earth tremble under them (this brought to my recollection a charge of cavalry). They stopt and gave us an opportunity to view them; among them were some very beautiful bays, blacks, and greys, and indeed of all colours. We fired at a black horse, with an idea of creasing him, but did not succeed; they flourished round and returned again to see us, when we returned to camp."[28]

Creasing, a practice described by Gregg as "a cruel expedient," was one method used to capture mustangs. It required expert shooting skill. A mustang was shot along the crease in his upper neck (above the cervical vertebrae) where nerves are present. A clean shot knocked him unconscious or temporarily stunned him so that he could be captured. A poorly aimed shot, on the other hand, would fracture a vertebra and kill the horse instantly, paralyze him, or cause a slow, painful death. If the aim was perfect, the wound would generally heal, although a hole made by a rifle ball was sometimes evident at the root of the horse's mane. European Americans adopted creasing upon their arrival on the frontier, often referring to it as "nicking." They also borrowed a variety of other methods used by the Spanish, Mexicans, and Indians to capture larger numbers of wild horses. One place on the Trail where hunters would wait to try to crease mustangs was called Pretty Encampment, a campground on the Mountain Route, where a stream ran through a grove of cottonwoods. This was east of Bent's Fort, where East Bridge Creek ran into the Arkansas River, in present-day Hamilton County, Kansas, not far from the present Colorado border.[29] Huge numbers of mustangs were found in this region of the Plains.

Pike and a few friends tried an "experiment" in which they used six of their fastest horses, equipped with ropes, to noose some wild horses that "stood until [the riders] came within forty yards of them, neighing and whinnowing, when the chase began, which we continued about two miles without success. Two of our horses ran up with them; we could not take them. . . . I have since laughed at our folly, for taking the wild horses, in that manner, is scarcely ever attempted."[30]

Pike mentioned noticing quantities of horse dung on the ground in some places. Mounds of freshly topped manure were the main mustang sign, showing that they were or had been present. Dobie remarked that celebrated author Washington Irving was "too elegant to mention the crudity" in his account of mustangs observed during his travel on the Santa Fe Trail in 1832, later published in *A Tour on the Prairies* (1850). Irving's travel companion, Henry Leavitt Ellsworth, was not so disinclined and noted "pyramids of manure often two or three feet high."[31] However, Washington Irving did marvel at the sight of a wild horse, "with ample mane and tail streaming in the wind," as it galloped away after haughtily gazing at his party. "It was the first time I had ever seen a horse scouring his native wilderness in all the pride and freedom of his nature. How different from the poor, mutilated, harnessed, checked, reined-up victim of luxury, caprice, and avarice, in our cities."[32]

Some of the earliest travelers on "the Road" wrote about seeing wild horses. For instance, William Becknell saw them in September 1821 during the inaugural journey over the soon-to-be-famous "highway of commerce." George Sibley, leader of the first survey of the Trail, recorded in his diary on September 3, 1825, in the vicinity of the Arkansas River and Clear Creek: "I saw six Wild Horses today on the high Prairie & shot at them, but without effect."[33] In 1831, Bostonian Albert Pike, amazed at their numbers, jotted in his journal: "It seems astonishing that from the few horses introduced so short a time since into America by the Spaniards, there should now be such immense herds in the prairies, and in the possession of the Aboriginals."[34] The sight made a lasting impression on him, and he later provided a glimpse for his readers: "Imagine yourself, kind reader, standing in a plain to which your eye can see no bounds. Not a tree, not a bush, not a shrub, not a tall weed lifts its head above the barren grandeur of the desert. . . . Imagine then . . . a herd of wild horses feeding in the distance or hurrying away from the hateful smell of man, with their manes floating, and a trampling like thunder."[35]

Health-seeker James Ross Larkin, member of an 1856 caravan led by William Bent, builder of Bent's Fort and brother of Charles Bent, saw a number of Cheyennes on their handsome horses, "making quite a hubbub." Some of them came to their camp to visit with Bent, who treated the visitors to coffee and bread. Larkin was very impressed by an Indian pony with its hooves and legs protected: "One of the Cheyennes has a novel set of shoes for his horse consisting of buffalo skins cut in round pieces about eight inches diameter [and] tied on with strings, the leather being placed over the bottom of the foot and tied around the legs."[36]

Marion Russell saw the wild horses during all of her five trips over the Trail. She described in her memoirs an unforgettable scene remembered from her trip made in 1860 when she was fifteen years old: "The buffalo were still numerous. Sometimes we had to take pains to avoid them. The country was so level that we could see for miles in all directions and the sun seemed to come up or go down like a great yellow disk right into or out of the earth. Sometimes we heard a noise like thunder and then a great herd of wild horses would swoop past us."[37]

From 1866 to 1867, Eveline M. Alexander traveled from New York to Fort Smith in Arkansas, and then to Fort Union with her husband, Captain Andrew J. Alexander of the Third Cavalry (he was promoted to major in the Eighth Cavalry in 1866). Frequently riding her own horse along the way, Eveline encountered a lone mustang stallion while she waited for the column to catch up: "As I was waiting quietly on Zaidee (her horse), a fine black horse about fourteen hands high came trotting through the woods. It approached within fifty yards of me and then threw up its head and galloped off, mane and tail flying. I thought at the time it was a horse from the first column that had been lost and found its way back here, but could not see any brand on him. . . . The guide said it was a wild horse, there being a large drove in the vicinity."[38]

Later, the company headed north from Fort Union, following the Mountain Route over Raton Pass to Fort Stevens in Colorado. Eveline and Andrew took a detour from "the Road" to see Rayado Ranch, owned by Lucien Maxwell of New Mexico, who sold his huge land grant a few years later in 1870. She also saw a Ute couple riding on colorful horses: "Their costume exceeded anything I have seen yet. The [young woman] had her hair divided into two tails, which were wound round and round with strings of small beads. Both her eyes and one of her cheeks were painted with vermillion [a bright red pigment]. She was dressed . . . in doeskin [and] her horse's bridle was covered with little bells, which jingled as she rode along, and she had a very pretty saddle cloth woven with bright colors and with a tassel hanging from each corner."[39]

Many Indians of the Plains cherished their ponies and horses, decorating them as colorfully as they decorated themselves. When the Indians were removed to reservations, they were forced to leave their beloved ponies and horses behind, to be taken, killed, or slaughtered by others. "Horses were to the Plains Indians what gold was to the whites," stated George Bent in his memoirs, a series of letters written in 1918.[40] The son of William Bent and Owl Woman, daughter of White Thunder, high priest of the Southern Cheyennes, George recalled that the largest of the wild herds were on the north side of the Arkansas in the area stretching from westernmost Kansas to east of Colorado Springs. The Cheyennes, Arapahos, and other northern tribes caught large numbers of mustangs in this area. The Plains Indians were particularly intent upon procuring more and more horses, although their herds were already huge. He remembered seeing the herds of the Comanches, Kiowas, and Apaches, who were camped along the Arkansas near his father's fort (Bent's Old Fort), grazing along the river for fifty miles.

George Bent explained how the Cheyennes and Arapahos hunted mustangs. The hunts were usually held in the early spring when mustangs were weak from the long hard winter and in poor condition for a hard run. The hunters set out from their camps on the north side of the Arkansas with their best horses and a number of old, gentle mares, while the scouts went ahead to find where the large mustang herds were ranging. The hunters kept out of sight in the rolling hills, staying upwind. When a promising herd was sighted, the scouts drew as near as possible and began spreading out to surround the mustangs. A scout on a very fast horse rode toward the herd, lying flat on his horse's back until he came close to it. Then, he quickly sat up and charged into the herd, making the mustangs flee in all directions.

The hunters rode out from behind the hills, each one picking out a mustang to run down, generally a young horse that would make the best mount for hunting buffalo. Each hunter held a long slender pole with the noose of his lasso fastened to the end. He rode up alongside a mustang and quickly slipped the noose over the mustang's head. Once the horse was choked down, the hunter put a rawhide halter on his head. Sometimes two or three men worked together and were able to catch several mustangs in a single run.

The mares now played their important role in the hunt. The halter on the captured mustang's head was tied close up to a mare's tail. This was called "tailing the mustang." The mustang would try to escape by jerking the mare about, but soon settled down and followed close behind her. After the hunt was completed, the mares were led back to camp with mustangs tied up close behind them. Upon arrival, the hunters picketed the mustangs near their lodges. The mustangs' heads were still tied to the mares' tails, and their feet were hobbled to keep them from kicking or breaking away. As they grew accustomed to seeing people moving about, the hunters would gentle them by touching and rubbing them. Then, the mare was untied and led around the camp with mustangs still tied to her tail. They would follow her without much trouble. Next, a buffalo robe was placed on a mustang's back. Once the horse was accustomed to the weight, a man would mount and sit on the mustang's back, repeating this many times until the mustang could be untied and led around with the mare alongside. After that, the Indian could generally ride the mustang without trouble.[41] This long process required much patience and skill.

When Lewis Garrard visited Cheyenne villages, he was intrigued by the way horses belonging to a lodge would stay close together. Each lodge had its own band of horses, which presented a strange sight. He described the scene: "Eighteen or more bands close to each other, walking along, but not mixing; each band following a favorite mare, perchance a woebegone, scrawny mule. . . . The sight of the different colored horses was gratifying in the extreme to my unaccustomed senses."[42]

Dobie declared that frontiersmen were "wrathy to kill" (their own words), and "shot mustangs for pleasure, just as they shot buffalo, jack rabbits, herons, and every other form of animal life within range."[43] Gregg made a similar observation: "The same barbarous propensity that was exhibited by the hunters toward the buffalo was also exhibited toward the wild horses. . . . Most persons appear unable to restrain this wonton inclination to take life, when a mustang approaches within rifle-shot. Many a stately steed thus falls a victim to the cruelty of man."[44]

———

With the arrival of the railroad in Santa Fe in 1880, the renowned Trail passed into history. The growth of the railroad and the public's increasing acceptance of the automobile in the early 1900s brought to a close the period known as the Age of Horse Culture. As the frontier and open range dwindled, the mustangs were increasingly considered a nuisance and competitors of domesticated stock for grass, forage, and water. Thousands were rounded up and sent off to war as cavalry mounts, to domestic and foreign markets to provide meat for humans and dogs, and to ranches and farms as work horses. Thousands more were killed.

Today, roaming herds and bands of mustangs are found in remote areas on public lands in about ten states of the American West, the largest population living in Nevada. The Pryor Mountain Wild Horse Refuge on the Wyoming-Montana border is one of these areas. Of the five states through which the Santa Fe Trail crossed, only New Mexico and Colorado are reported to have small bands of mustangs living in areas far from the ruts of the old Trail.

In 1971, heightened public concern over the brutal treatment of mustangs and their decreasing numbers gave impetus to the passage of the Wild Free-Roaming Horse and Burro Act by the U.S. Congress. Mustangs were also designated as a national heritage species, living symbols of the history and pioneer spirit of the West. The Adopt a Horse or Burro Program was established in 1973 and continues to place mustangs with people who give them shelter and care.

Places names scattered over the Great Plains and along the routes of the Santa Fe Trail provide us with reminders of those days when mustangs roamed wild and free. Among those places are Wild Horse Lake and Wild Horse Draw in western Kansas; the town of Wild Horse and Wild Horse Creek in Colorado; Wild Horse Mountain and Horse Pen Creek in Oklahoma; and Mustang Creek and Caballo (Horse) Mountain in New Mexico. These and other names call back those vivid scenes that Trail travelers saw and, fortunately for us, that some recorded. Through them, we are able to create our own visions of those magnificent, colorful, carefree mustangs "gamboling and curvetting" over the grassy plains.

Part II

Domestic Animals on the Santa Fe Trail

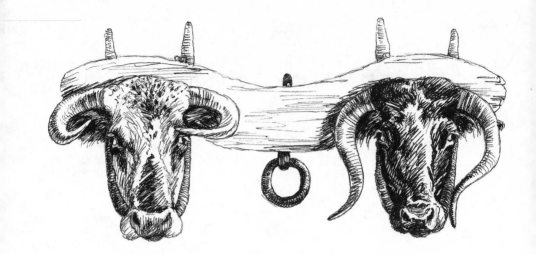

Chapter Ten

Oxen ✓ 7-3-16

An ox is a mammal and a member of the Bovidae family, to which all bovines, or cattle (*Bos taurus*), belong. The term "cattle" comprises cows, bulls, and steers. Not a separate or special breed of bovines, oxen are cattle that have been trained to work as draft animals. Kansan James Francis Riley explained this important point about oxen in his "Recollections," in which he recounted preparations for the start of his first trip over the Santa Fe Trail in a wagon train departing from Independence, Missouri, in 1859: "In eight or ten days we received our cattle which had been purchased in southwest Missouri. Now our work commenced. We had to build a corral and branding stall and brand our three hundred and fifty head of cattle. I say cattle for a great many of them had never been broke to work and, of course, they were not oxen until they were broke to work."[1]

Oxen are often thought of as a pair or yoke of two, also called a span, because they have long been paired in teams and are generally depicted in this manner in photographs, paintings, and other illustrations. Their physical characteristics include muscular bodies, sturdy legs, cloven (split) hooves, long tails, and usually horns situated on the sides of their heads. Horns were necessary for the Spanish horn yoke, but not for the American neck yoke. A frequently asked question is whether a team of two yoked oxen can pull more than double the combined load that two separate oxen are able to pull. The answer is "yes," with the qualification "if the conditions, training, and capability of the teamster are optimum."[2] Thus, the duties of the teamster, also called ox drover or bullwhacker, were very important to the wagon trains on the Santa Fe Trail.

Cattle trained and used in pulling wagons and other types of heavy work have generally been adult, castrated males (steers), although adult females (cows) and bulls (intact males) have also been used throughout the long history of draft animals. Steers and bulls were primarily used on the Trail for freighting. Some emigrant families used the family milk cows to pull their wagons to a new home. While almost any breed of cattle can be trained as oxen, some breeds have been favored, in particular the long-horned Red Devon, Durham, Galloway, and Ayrshire. It was not unusual for teams to consist of wiry, wily, and wild Texas range steers.

Oxen have transported cumbersome carts and wagons, hauled logs, threshed grain, pulled plows, and performed countless other burdensome tasks, making the labor of human beings around the world much easier. Spaniards were the first to bring cattle to New Mexico, more than four hundred years ago. Oxen had been used as draft animals in Spain since the Roman Empire. Historian Marc Simmons has written about Juan de Oñate and colonists, who brought cattle with them to New Mexico in 1598. Thereafter, oxen would serve on the trails and in the fields of Hispanic New Mexicans and Pueblo Indians until the twentieth century.[3] Oxen are still used as draft animals in other areas of the world. $\vee$

The wooden Spanish horn yoke was fitted across the top of the oxen's heads behind their horns and tied to the horns with rawhide straps. The American yoke, called a neck yoke, consisted of a pair of hardwood U-shaped oxbows fitted around each ox's neck. The upper ends of the "U" fitted into holes made in the yoke. Wood pins or metal keys secured the bows in the yoke, which had an iron ring hanging from its center that was used in the hitching of the oxen to a wagon. The design of the neck yoke was more effective because it allowed oxen to pull the loads with their shoulders rather than their heads and necks. This type of yoke is used as the logo of the national Santa Fe Trail Association.

Horses and mules were the only draft animals used by the traders during the initial years of the Santa Fe Trail. Starting in 1821, William Becknell and his small party of traders relied on horses, which were soon replaced by mules. It was through an experiment that oxen came to be used as draft animals on the Trail. Major Bennet C. Riley (later Colonel and Brevet Major General Riley) experimented with oxen when he commanded the first military escort providing protection for wagon trains crossing the prairie in 1829. He purchased seventy-two oxen to pull supply wagons for his infantry battalion because oxen were less expensive than mules. The oxen cost about one-third as much as mules, and they could be eaten if necessary. In fact, Riley purchased the oxen for his supply and baggage train with commissary funds.

The escort left Fort Leavenworth (only a cantonment, or temporary military quarters at that time) in the spring of 1829 and joined the annual trade caravan on the Trail at Round Grove. The soldiers escorted the traders west to the boundary line between the United States and Mexico at the upper crossing of the Arkansas River (in present Kearny County, Kansas). After crossing the Arkansas, the caravan headed for Santa Fe on its own, while the escort, which did not have permission to enter Mexico, waited near the river for the caravan's return from Santa Fe and escorted it back to Fort Leavenworth. Riley thought the oxen's performance was nearly equal to that of mules. Captain Philip St. George Cooke of the Dragoons, who accompanied the command, noted in his journal: "Up to this time, traders had always used horses or mules. Our oxen were an experiment, and it succeeded admirably, they even did better when water was very scarce, which is an important consideration."[4]

Caravan captain Charles Bent borrowed oxen from Riley's command to pull a freight wagon to Santa Fe, testing to see how oxen would perform pulling a heavy load. Although the oxen were lost during the trip to Santa Fe, Bent was satisfied with their performance. He used oxen again to pull some wagons in 1830. From that time, oxen quickly became accepted as draft animals on the Santa Fe Trail. Because they cost less than mules, oxen were especially favored by large freighting companies, which required hundreds of draft animals.

As the smaller wagons used in the early years grew into huge, heavily laden freight wagons carrying three thousand and more pounds of trade goods, oxen became the favored draft animal. Each weighing as much as eighteen hundred to two thousand pounds, they had the power to pull those cumbersome loads over long distances. The customary number of yokes of oxen per wagon used on the Trail was six or eight. In many instances, such as when teams were doubled to pull wagons over obstacles, through deep sand, or out of muddy river bottoms, the number could reach twenty or more. Even with that number, they were thought to be better than mules. Veteran trader Josiah Gregg, counting in his *Commerce of the Prairies* the advantages of using oxen, mentions their "pulling heavier loads than the same number of mules, particularly through muddy or sandy places."[5]

Gregg, who had traded on the Trail since 1831, saw oxen become increasingly popular. "Upon an average, about half of the wagons in these expeditions have been drawn by oxen,"[6] he said in his book, which was published in 1844. By 1860 and throughout the last two decades of the Trail, until the railroad replaced it in 1880, oxen reigned supreme as the most preferred draft animal on the Trail. Mules were also used throughout the same period, but not in the numbers oxen reached. David K. Clapsaddle, Santa Fe Trail ambassador and author of numerous works about the Trail, has cited figures showing that "oxen outnumbered mules at the rate of more than six to one."[7] For example, a record from 1865 shows that 38,281 oxen were counted on the Trail, compared with 6,452 mules.[8]

The monetary savings in using oxen went far beyond their purchase price. The equipment used for them was less expensive than that required for mules. Oxen also cost less to feed, because they could subsist more easily on range or prairie grass.[9]

Ox drovers also thought oxen were less likely to be stampeded or taken by Indians. Although Josiah Gregg and others agreed that Indians had little or no interest in cattle or oxen, they did not agree about oxen when it came to a stampede, often called "a general scamper" in the Trail's early years. Gregg said that oxen "are decidedly the worst when once started."[10] Gregg told about a stampede beginning at midnight when the cattle and mules were shut up inside the corral made by the circling of the wagons. Except for the night guard, all of the people in the caravan were asleep. The panic, caused by a dog, quickly turned into a stampede: "Although the wagons were tightly bound together, wheel to wheel, with ropes and chains and several stretched across the gaps at the corners of the corral, the oxen soon burst their way out, and though mostly yoked in pairs, they went scampering over the plains. All attempts to stop them were vain."[11] In the early morning, oxen were found six or seven miles away, and all but half a dozen were recovered. The mules had been quickly retaken, and none were lost.

Well-trained oxen respond to verbal commands and to body language. Among frequently used spoken commands on the Trail and still used today are "gee" for a right turn and "haw" for a left turn. Oxen have been raised since colonial times in the United States, with training taking place over three or four years. In Trail days, there was little or no time for training of either oxen or the greenhorns who were employed to control them. Thus, men and oxen had to learn along their way to Santa Fe. James Francis Riley described such a situation: "We had in our outfit eight or ten hands who had just come out from Illinois and had never driven an ox and some that had never seen an ox team until they came here. To you that may seem a small matter, but to me it was quite a problem. For it is often easier to break in a wild team than to teach a green man. . .

"After all the hard work in getting everything ready was over, then comes the hardest job of all. That job is yoking up and matching up the teams, especially for the inexperienced hands and where one has a good many unbroken cattle. It usually takes the biggest part of a day and is very hard work. After all is ready we make our start. It is mostly herding for a day or two, but the main thing is to keep in the road and it doesn't matter so much how one does it, just so he keeps in the road. In a very short time your cattle will learn it."[12]

At the age of twenty-two, Franz Huning emigrated from Germany and arrived at St. Louis in 1849. He joined a wagon train at Fort Leavenworth as a bullwhacker. He recalled in his *Memoirs* about "getting his taste of the hardships of 'bullwhacking' in midwinter." The wagons headed out in October "towards the Great Plains, where we expected 'to see the elephant' [see amazing and wondrous things]."[13] Instead, the first day almost proved fatal to Huning, and one of his companions was fired as "utterly worthless" after only an hour on the road. Ten days later, after traveling in a manner Huning described as "rather irregular," they arrived at Council Grove. By that time, "green hands had learned the business of bullwhacking, could yoke up the oxen, fix their whips, make keys for the bows, etc.; and last, but not least, had learned to swear and had become quite proficient in the use of regular oaths of the profession they readily learned from the old teamsters in the train. They—the greenhorns—had found out very soon that they need not attempt to drive oxen without swearing. . . . When a team got stalled in a mud hole or in a sandbank, the profanity that was considered necessary to pull it out was terrible, and it sometimes appeared to me that it scared even the oxen."[14]

Along the way, the wagon master turned over to Huning a team usually driven by one of the regular teamsters, who had been confined to bed in his wagon because of worsening consumption (tuberculosis). He recounted the experience: "It [the team] was in fair order and gentle. The oxen understood their duty as well as their present driver, and I got along first rate. They knew their place in the train; I yoked each pair, gave them a slap on the back and they would find their proper place; I hooked on the chains and my team was ready to start, sometimes fully a quarter of an hour before most of the others were ready. The sick man still took an interest in his team; he would sometimes raise the wagon sheet to look out and give me some direction, or other, always cautioning me not to whip the oxen, and it really was seldom necessary, there was not a better team in the train."[15] Many oxen perished because of bitterly cold weather. Huning remembered the animals were "very poor" by the time they reached Las Vegas, New Mexico. When he rode into Santa Fe, his team had been reduced from six yokes to three and one-half yokes of very poor cattle. Huning would later start a mercantile business in Albuquerque and made more than forty trips on the Trail to Kansas City to purchase goods for it.

In 1866, Tom C. Cranmer published *Rules and Regulations, by Which to Conduct Wagon Trains (Drawn by Oxen on the Plains)* in Kansas City, Missouri, and this publication was reissued by the Wet/Dry Routes Chapter of the Santa Fe Trail Association in 2007. In the preface, the chapter's president, David K. Clapsaddle, explains that Cranmer's document was meant for "freighters of less magnitude than Russell, Majors and Waddell," because such large freighting companies hauling government contracts "regulated the work of their employees in an institutionalized manner . . . and issued printed instructions to their employees."[16]

Cranmer included instructions relating to the duties of the ox drover, the yoking and unyoking of oxen, and the doubling of teams. He also suggested the makeup of a train and its personnel, stating that he considered a night herder's duties in wagon trains drawn by oxen as "the most important of all train business."[17] Later in the book he explains his rationale: "Long experience has taught me that, if for the first few nights it should take the two regular night herders, together with the wagon master and his assistant, and half the regular teamsters to keep the cattle, it is still easier to keep them while you have them, than to hunt them when they are lost."[18]

Every wagon train had a loose herd of animals that followed at the very end of the caravan. This herd included extra cattle, oxen, mules, and horses, some of which would replace any animals that became lame or died. Names used for these loose herds on the Trail came from the original Spanish word *caballada,* referring to a herd of saddle horses. People on the Trail who were not familiar with Spanish turned this word into a variety of words and spellings. The following eight terms for the loose stock were found in the research for this chapter: cavallard, cavayado, cavieyard, cavvyard, cavvy, cavey yard, cavayard, and caviyard. Cranmer used two of these, caviyard and cavieyard, in his chapter titled "Duty of the Cavieyard Driver."[19] The reason for all of these words beginning with "cav" is because a "b" in Spanish, as in *caballada,* often sounds like a "v" to a person unfamiliar with Spanish.

Cranmer included the following advice on the yoking of oxen: "A teamster should never bring his whip into a corral to drive out [the oxen]; always, if necessary, use a small stick, making as little noise as possible. No loud cursing, swearing, or fighting [the] cattle should ever be allowed in the corral; if so, the cattle soon become gentle and quiet. . . . If men go into the corral cursing, and damning, and creating unnecessary tumult, the herd will be always in a stir, and they will have a separate hunt for every ox. A man never gentles his own team by fighting it unnecessarily, and invariably frightens others, and makes them unmanageable as well as his own. I would, therefore, most emphatically denounce the practice of beating oxen under all circumstances."[20]

Susan Shelby Magoffin, traveling in her carriage with her husband's wagon train in 1846, would have heartily agreed with Cranmer's admonishment concerning the treatment of oxen. She thought the swearing, profanity, and loud commotion was "disagreeable." On her first Sabbath on the plains, she noted in her diary how quiet it was: "I have scarcely heard an oath the whole day."[21] Since it was Sunday, the wagon train was in camp and not moving on the Trail. It seems, however, that the majority of ox drovers and bull-whackers accepted a great deal of swearing and whipping as absolutely necessary. In writing about the profanity directed toward oxen, Simmons tells about an early-day preacher advising his flock to refrain from using profane language, saying at no time was it excusable unless they were "whacking bulls."[22]

The freighting firm of Russell, Majors and Waddell required employees to sign a pledge not to swear or use profanity and gave each man a Bible to carry on the Trail. These men were known by others as the Bible Backs or the B.B.'s. George Vanderwalker recalled meeting the B.B.'s in a Russell, Majors and Waddell outfit on the rugged Raton Pass near Trinidad, Colorado: "The pass in those days was surely equal to the 'rocky road to Dublin,' and from the conversation of the 'B.B.s' and the language they were using toward their cattle, one would infer the men had lost the 'Word.'"[23]

———

Rejected for enlistment to fight in the Civil War because of his age and short stature, George E. Vanderwalker, a thirteen-year-old from Michigan, signed on for his first experience in "whacking" in 1864. He arrived at Diamond Spring, where he was employed by the Stewart Stemens Company of Council Grove. He told in his "Reminiscences" about his instruction in "the art of how to handle a wagon with a live end to it." The wagon master and his assistant taught Vanderwalker the proper manner of carrying an ox yoke and bow in yoking the cattle in preparation for hitching them to the wagons. He was given a whip with a lash about sixteen feet long with a "popper" (a whip cracker that made a loud sound like a pistol shot) and fastened to a whip stock [the handle] eighteen inches long by a buckskin thong. He described his whip: "This instrument of torture required an almost constant every day manipulation by me during my first two hundred miles of the trip before I became proficient enough in handling it to prevent its going about my neck and hanging me."[24]

When a wagon train was on the road, the wagon master, riding his mule, was up front to check their way ahead. The assistant to the wagon master was on the near side of the train about the center, so that he could see what was happening at both ends of the train. The mounted extras, whose duty it was to carry out all orders given by the wagon master or his assistant, would generally be on the sides. The ox drover's place was on the ground near the heads of his "wheelers," the yoke of oxen closest to the wagon, except in poor weather or road conditions when he had to be all along his team. In the middle of the team were the "swings," spans or yokes of "green" oxen placed where they could do the least damage. At the front of these spans were the leaders, well-trained oxen who were lighter than the wheelers and could set a faster pace for the team.[25]

According to some experienced travelers, the pace of the oxen was more fluid than that of horses and mules, allowing a more comfortable ride. Marion Sloan Russell was in a position to make comparisons; she traveled the Trail five times, beginning at the age of seven in 1852. "Mules draw a wagon a bit more gently than horses," she recalled, "but oxen are best of all. 'Tis true they walk slowly, but there is a rhythm in their walking that sways the great wagons gently."[26] Marion and her mother placed their bedding in the back of a freight wagon packed with boxes and bales for Fort Union in New Mexico. The bedding made a comfortable place to rest and sleep, even though there were bumps along the way.

"In the days of the Old West," writes Simmons, "wagon freighters often sat around the campfire arguing the merits of oxen versus mules, the way men today will argue the advantages of one brand or model of car or truck over another. . . . The oxen vs. mules debate was not easily settled. Teamsters accustomed to driving a yoke of oxen would swear by their draft animals and claim that the mule was an inferior beast. Mule skinners, on the other hand, were just as convinced that their animals were perfect for long-distance hauling and that the dumb ox could not compete. In truth, the ox and the mule each had his selling points, his special qualifications. . . . But both also had unique disadvantages that worked against them."[27]

Among the drawbacks of using oxen was that they did not tolerate very hot weather. This stemmed from their not having sweat glands, which mules and horses do have. When oxen became overheated, it was necessary to stop and rest along the Trail. Resting in a shaded area was helpful, but that was often not available, especially on the Cimarron Route, the Dry Route. Because of this, teamsters drove the wagon trains in the early morning, stopped during midday for rest, and continued their way later in the afternoon and evening. Oxen were also more susceptible than mules to a variety of infectious diseases known by the collective term "murrain." One infectious disease, called the Spanish fever, spread along the Trail in the 1850s, killing large numbers of oxen. Another disease, called the Texas fever, struck in the 1870s. Oxen were also susceptible to alkali poisoning, especially in very dry years when the water and forage had a higher alkali content.[28]

While oxen and mules could get along on sandy ground without shoes, oxen more often than mules needed to be shod to protect their feet. They have sensitive, tender hooves, which became very sore, especially on rough or rocky terrain, thus debilitating them. When mules had to be shod, it was relatively easy to nail a shoe on a mule's hoof anywhere by picking up and holding a foot while the mule stood still. Oxen posed a problem because they are unable to stand on three legs. Blacksmiths in such places as Council Grove generally had a set of wooden stocks, a mechanical device used to hold an ox and turn him on his side. This allowed nailing the iron shoes on the ox's hooves. Since oxen have split hooves, each foot requires a set of two shoes. When stocks were not available, a narrow trench was dug in the ground and the ox was turned on his side or back with the use of ropes. Shoes were also made out of rawhide, fitted around each foot, and securely tied. These "moccasins" worked, but they did not last long in bad weather or muddy conditions.

Old-time traders and freighters who crossed the plains established the Old Plainsmen's Association in 1909 and held annual reunions to 1917. Alexander Majors of Russell, Majors and Waddell, spoke at one of those reunions: "I remember once of timing my teamsters when they commenced to yoke their teams after the cattle had been driven into their corral and allowed to stand long enough to become quiet. I gave the word to the men to commence yoking, and held my watch in my hand while they did so, and in sixteen minutes from the time they commenced, each man had yoked six pairs of oxen and had them hitched to their wagons ready to move."[29] Another former freighter, Charles Raber, who settled in Westport, Missouri (now a historic neighborhood in Kansas City) recalled some of his experiences in partnering in a freighting business in the 1860s. He told the assembly that one of their oxen had traveled more than ten thousand miles on the Santa Fe Trail.

On the Santa Fe Trail and other overland trails, oxen provided the greatest amount of muscle power, as well as milk, meat for food, and leather. Thousands died along the way, their bleached bones scattered across the plains

among those of mules, horses, and other domestic animals that partici-
pated in and contributed to our nation's growth and development. Surely,
many of the people who carried their Bibles with them along the Santa Fe
Trail, including the "Bible Backs," were familiar with these verses from the
book of Matthew in the New Testament: "Come unto me, all ye that labour
and are heavy laden, and I will give you rest. Take my yoke upon you, and
learn of me; for I am meek and lowly in heart; and ye shall find rest unto
your souls. For my yoke is easy, and my burden is light."[30] Quite possibly,
some of those people thought of the oxen drawing the wagons or carriages,
of their suffering, and gave thanks for them.

Chapter Eleven

Mules

A number of historians and others have extolled the contributions of the mule. Historian Max L. Moorhead wrote that the mule has too often been overlooked, and unfairly: "Much has been written about the Spanish horse in the conquest of the New World . . . but the unsung hero of transportation in the Southwest was unquestionably the Spanish mules."[1] Marc Simmons stated in a more recent essay that "without the mule, the history of the Southwest might have run a different course."[2] Texas historian Floyd F. Ewing, Jr., would have emphatically agreed. This "poorly regarded hybrid," he wrote in an essay devoted to the mule's historic role, was fundamental to the development of the Southwest. "Between 1820 and 1860 the role the mule played in the great task of subduing virgin lands and fashioning channels of trade and commerce in the Southwest was just as spectacular and just as important as that of the mustang or longhorn."[3] Ewing also credited the Santa Fe Trail for creating the first great demand for mules and being a major source of these hardworking animals.

A mule is a domesticated hybrid resulting from the mating of two species of the equine family, the donkey (*Equus assinus)* and the horse (*Equus caballus*). The offspring has the characteristics of both parents. A cross between a donkey stallion, called a jack, and a horse mare produces a mule, while a cross between a stallion horse and a donkey mare, called jennet or jenny, results in a hinny. Hinnies are generally classified under the general term, mule. A mule or hinny may be a male (horse mule or horse hinny) or a female (mare mule or mare hinny). Sometimes horse mules (males) are called Johns, and the mare mules are called Mollies. All are sterile and cannot reproduce, except in very rare instances (one in one million) when a mare mule has a foal. Since mules are not a species, an approximation of a Latin name that has been used for the mule, according to the American Donkey and Mule Society, is *Equus mulis* and a correct scientific classification is Hybrid (*Equus caballus x Equus assinus*).[4]

Mules, or mule-like hybrids, have been in existence since ancient times. They have been used as work animals around the world and valued for their hardiness, strength, sure-footedness, and resistance to disease. In the 1600s, long trains of pack mules made their way from Mexico City and northern Mexico to the new settlements the Spaniards established in the Rio Grande Valley. Mules were used in the eastern United States since colonial times. In fact, George Washington was a respected livestock breeder and began a breeding program to improve the mule. He received a splendid jack, a donkey stallion named Royal Gift, from the king of Spain for his program. Mules were present along the Santa Fe Trail from its earliest years.

On September 1, 1821, the year Missouri became a state, a small band of men with pack horses loaded with trade goods left the village of Franklin in Howard County. The leader was merchant William Becknell. The men headed south and west to Mexico and, in particular, to Santa Fe. Along their way, Mexican soldiers informed them that Mexico had finally won its independence from Spain, and traders from the United States would be welcomed. Continuing to Santa Fe, Becknell and the men with him reached their destination on November 16, two and one-half months and more than eight hundred miles after setting out. Upon their return to Franklin on January 29, 1822, a glittering shower of Spanish silver coins, falling onto the ground from the traders' opened packs, excited the crowd formed around them. Becknell's success opened regular trade with Santa Fe, which was part of the Republic of Mexico's far north. Thus 1821 was designated the year of the opening of the Santa Fe Trail, which historian Marc Simmons describes as "first and last a highway of commerce."[5] Trade and travel over the Santa Fe Trail's routes flourished until the railroad replaced it in 1880.

When spring arrived in 1822, Becknell led another expedition to Santa Fe with at least three wagons. Mountain man Jacob Fowler, on his way down the Arkansas River, told of seeing the tracks made by their wagons near his camp—an astonishing sight at that time and place. This trading party carried $3,000 worth of trade goods and made a profit of 2,000 percent on their investment.[6] Two other expeditions were made in 1822. One of those, headed by Colonel Benjamin Cooper, a member of the Cooper family of the Boone's Lick area near Franklin, used pack mules and returned from Santa Fe with more than a hundred mules and other trade items.

The other expedition of 1822 was led by James Baird and Samuel Chambers. A heavy snowstorm forced them to stop along the Arkansas near the Cimarron Crossing. They had to stay through the winter, and their pack animals perished. They dug deep holes in the earth to store their goods for safekeeping. This place became known on the Santa Fe Trail as "The Caches," marked by a monument west of present-day Dodge City, Kansas. The men left the area for Taos and returned with fresh animals to continue on their way to Santa Fe.

In 1823, the only recorded expedition to Santa Fe was led by Major Stephen Cooper, who had accompanied his uncle Benjamin the year before. The major organized a company of about thirty men with stock in trade consisting primarily of dry goods. On the Little Arkansas, Indians ran off all but six of the horses, and Cooper was forced to return to Missouri to obtain more animals. When they returned to Missouri in October 1823, their trade items included four hundred jacks (donkey stallions), jennets (donkey mares), and mules. Robert Duffus speculates in *The Santa Fe Trail* that Cooper's expedition was history in the making: "Their four-legged booty was apparently the beginning of the now world-renowned Missouri mule. This notorious beast was a New Mexican product. He invaded Missouri from the west, filling a need which the rush of settlement into the river country was just beginning to create. From 1823 onward, the mule formed 'a conspicuous article of commerce.'"[7] Before this time, records made no mention of mules in the state that later became nationally and internationally famous for them.

In May 1824, a large caravan left Howard County, Missouri, where Franklin was located, with eighty-one traders, 156 mules and horses, twenty-five wagons, and a small piece of field artillery. Led by Alexander Le Grande, the company also included Augustus Storrs, who became U.S. consul at Santa Fe the following year, and Meredith Miles Marmaduke, the future governor of Missouri. The caravan carried $35,000 in trade goods. The traders arrived in Santa Fe on July 28 and were back in Franklin on September 24. The trip was declared successful in every way.[8]

In 1825, when $65,000 to $85,000 in goods went over the Trail, Senator Thomas Hart Benton of Missouri stated: "The New Mexican trade has grown up to be a new and regular branch of interior commerce, profitable to those engaged in it, valuable to the country from the articles it carried out and for the silver, furs, and mules it brought back, and well suited to the care and protection of our government."[9] For the Trail's first fifteen years, furs, especially beaver and otter skins, and buffalo robes were important. Most caravans heading back to Missouri carried at least some furs. The mules and precious metals, however, remained important for a much longer time.

————

Trouble with the Indians on the plains increased in the late 1820s. In 1827 the Pawnee attacked a company returning east and made off with a hundred head of mules and other animals. A total of eight hundred mules were brought over the Trail that year. In the autumn of 1828, a caravan was returning with a thousand head of mules and horses and were near the present Oklahoma–New Mexico boundary (northeastern New Mexico) when Indians killed young traders Daniel Monroe and Robert McNees. This killing occurred at the crossing of the North Canadian River, called Corrumpa Creek at that time and renamed McNees Crossing. This incident, in particular, fueled fires of retribution that lasted for generations. All of the thousand head of livestock were lost before reaching home. After this great loss, traders demanded protection. For them, losing mules was the same as losing silver; mules were valuable in Missouri. Indians valued mules, too, preferring them over coin as booty and over oxen, in which they showed little interest. They used mules for riding, carrying loads, and providing food when buffalo meat was not available.

Duffus commented about the animals used early on in the opening of the West and along the Santa Fe Trail: "First came the patient pack-horse, the faithful companion of the fur trader in every western journey. The pack-horse could go almost anywhere a man could go. But he was not so good for carrying heavy burdens on a long and relatively easy trail as was the mule. Besides, the mule was abundant in New Mexico. He had his faults, to be sure. Sometimes his energy would display itself at inopportune moments, as when, after having walked four or five hundred miles, he would suddenly 'take fright from a profile view of his own shadow and run like an antelope of the plains.' But very often the mule outfits would come into Santa Fe, after eight hundred miles or more of travel, in pretty good condition."[10]

By the 1830s, a change was beginning to take shape in the Santa Fe trade; it was becoming an occupation for businessmen. This change was under way when trader Josiah Gregg made his first of four trading trips over the Trail from 1831 to 1840. With his 1844 classic, *Commerce of the Prairies,* Gregg became one of the first to write about mules and their popularity and usefulness. He thought mules had a number of merits, including being quick to sense danger. Their "vigorously twitching ears," for instance, let guards know that something was awry or that Indians were nearby. But he also mentioned the mules' "pretentious ways," probably a tactful way of saying that mules could be troublesome.[11]

Gregg was greatly impressed with the Mexican *arrieros,* the muleteers who trained and packed the mules on the Trail and in New Mexico. He described in considerable detail the muleteers' skills in handling mules and their methods of packing. These men knew their business, gave their lives to their profession, and were devoted to the mules. An expertly packed mule with a heavy load of unwieldy articles weighing as much as three or four hundred pounds could travel all day and into the evening, and be ready to go the following morning. The techniques of the *arrieros* were adopted early by the U.S. Army, which used mules for packing, riding, pulling wagons and other forms of transportation, carrying mail, moving artillery and munitions, and numerous other purposes. Army mules wore the historic brand, U.S., on their left hips and served many decades. The last two military mule units were deactivated at Fort Carson, Colorado, in December 1956.

Simmons gained knowledge of *arrieros* and their packing expertise, muleteering, and horseshoeing through his research and study in the Southwest, Mexico, and Spain. He has written about those muleteers: "The work of the *arrieros* was highly specialized, a fact that also contributed to their clannish spirit. Years of practice were required to learn the many skills of the business. . . . Most of the rich lore surrounding the packing profession has been lost. But one of the old sayings that does survive expresses the true spirit of those plucky muleteers: 'Better to be an *arriero* than to be rich.'"[12]

———

When it came to overland transport of goods, among the advantages of mules over horses and oxen were their endurance and hardiness. Their feet are denser and tougher, often allowing them to make a full trip on their hooves without being shod. Oxen, on the other hand, generally needed iron shoes. Mules are sure-footed and good mountain climbers, allowing them to go places where horses and oxen could not go, such as rough terrain, rocky country, lava beds, and long, arid stretches of land. They also held up better under adverse conditions, such as heat, lack of water, and sparse forage. They avoided eating poisonous weeds, overeating, and drinking too much cold water on a hot day. They could travel great distances with considerable speed. Simmons has told of a company of scouts with a mule pack train marching eighty-five miles in twelve hours under the blazing New Mexico sun. "That feat becomes all the more remarkable when we recall that the average distance covered by a horseback rider in a day was usually only twenty-five to thirty miles."[13] Mules' advantages were recognized by many, including riders who learned of their value as saddle stock.

Among the disadvantages of mules was their purchase price. They cost more than oxen, an important consideration for traders, merchants, freighters, riders, and others. The price could be as high as $100 per head in the frontier towns, while oxen cost about $25. Also, many men on the Trail thought mules were much more stubborn and cantankerous than oxen or horses, and they were especially known for their ability to kick. They could cause serious injury, just as horses, oxen, and other animals, domesticated or wild, also caused injury to people on the Trail.

Many of the behavioral problems experienced with mules were a direct result of improper handling. Inexperienced traders and greenhorns frequently were in a rush to prepare the animals for the journey over the Trail and decided, wrongly, that they could cut corners by beating, lashing, and whipping the animals. Simmons has commented: "Army training manuals explained that the young mule kicked because he was afraid of man. Training and kind treatment could cure that. But many handlers lacked patience and so they resorted to the whip, which in the end produced a confirmed kicker. Not surprisingly, the mule has inspired the use of more strong language than any other animal."[14]

Susan Shelby Magoffin noted that strong language with disapproval in her diary entry of Thursday, June 11, 1846, which was her first day on the Trail, starting out after breakfast at seven o'clock. The part of the caravan in which she was riding in her personal carriage began heading toward the prairie to meet the rest of the caravan, which was waiting ahead for them to arrive. Susan recalled: "Now the Prairie life begins! We soon left the 'settlements' this morning. Our mules travel well and we jog[g]ed on at a rapid pace till 10 o'clock, when we came up to the waggons. They were encamped just at the edge of the last woods. As we proceeded from this thick wood of oaks and scrubby underbrush, my eyes were unable to satiate their longing for a sight of the wide spreading plains. . . . All our waggons were here, and those of two or three others of the traders. The animals made an extensive show indeed. Mules and oxen scattered in all directions. The teamsters were just 'catching up,' and the cracking of whips, lowing of cattle, braying of the mules, whooping and hallowing of the men was a novel sight. . . . It is disagreeable to hear so much swearing; the animals are unruly 'tis true and worries the patience of their drivers, but I scarcely think they need be so profane."[15]

Susan continued: "And the mules I believe are worse, for they kick and run so much faster. It is a common circumstance for a mule (when first brought into service) while they are hitching him in, to break away with chains and harness all on, and run for a half hour or more with two or three horsemen at his heels endeavouring to stop him, or at least to keep him from running among the other stock. I saw a scamper while I sat in the carriage today. One of the mules scampered off, turning the heads of the whole collection nearly by the rattling of the chains. After a fine race, one of his pursuers succeeded in catching the bridle, when the stubborn animal refused to [follow] and in defiance of all the man could do, [the mule] walked backwards all the way to camp leading his capturer instead of being led."[16]

Among other Trail travelers writing of the trouble with animals taking off in every direction whenever the opportunity arose was William James Hinchey, an Irishman born in 1829 who later became a popular nineteenth-century St. Louis portrait artist. Traveling the Trail in 1854 and 1855, he was part of the entourage of Bishop Jean Baptiste Lamy, who later became the archbishop of Santa Fe.

Hinchey kept a journal of his experiences and drew in it many sketches of events and sights along the way. While writing in his journal on a September Sunday in 1854, he had to stop and join a chase after mules and other animals, including the white mare that caused it. Upon returning to his tent, Hinchey resumed writing: "Confound the mules! Those vile beasts who keep us all employed the whole day and half the night—they can't come and eat quietly out of their long carved trough; they must be kicking and screeching at one another. Well, now we're all scattered here and there trying to hunt them up. . . . She [the white mare] has done all the damage being the first to start off and still keeping the lead of the whole party. All the animals [have] taken allarum [alarm] and not wishing to be too easily entrapped have set off full speed, passing all the party, priests, deacons, hunters, Mexicans, French, Americans, Spanish, Irish and all, and away with a train of forty quadrupeds—horses, mules, jennets, ponies—and after them a train of bipeds consisting of all the different nations above mentioned in all their various tongues, and scattering the animals they want to catch, in every direction."[17]

The mare was finally lassoed and some of the other animals captured. As these were being led back to the encampment, the remaining animals fell in line behind them, "walking along with their captive fellows as though they were above deserting them in the moment of need and should go with them to console them in their imprisonment." Hinchey ended his journal entry: "Now then, I too am returned to camp determined no more to go mule hunting on foot for the exertion is too great."[18]

Although Hinchey wrote about the "chase" in a light, humorous way, a stampede was a very serious matter and sometimes caused disastrous results for men and animals. Almost anything could set off a stampede: a sudden noise, the howling of a wolf, raiding Indians, a prairie fire, a passing buffalo herd, a blizzard, or thunderstorms and lightning. The stampede could occur in the light of day or the darkness of night, an especially terrible time for it to happen. Cautious wagon masters and mule skinners kept the herders and guards saddled, armed, and ready for the slightest possibility of a stampede or runaways. Mules, for example, had a habit of running off with buffalo herds. Stanley Vestal described such an occurrence: "The beat of a thousand hooves obliterated their trail, they could seldom be recovered. Even in the first stages of the trip, mules were hard to follow, harder to catch, because of the patches of timber, numerous creeks, constant rains, and the speed with which they hastened back to the settlements."[19]

While others along the Trail had a strong dislike for mules, James Josiah Webb, a Connecticut storekeeper and merchant involved in the Santa Fe trade from 1844 to 1847, loved and admired his mule. He called her Dolly Spanker (in those times, a "spanker" was someone or something that was very special and highly regarded). She had been a very good Trail companion, reliable, and swift and smart in chasing down buffalo. Later in his life, Webb wrote a tribute to her: "This is a long story about a mule, but Dolly with all her naughtiness was an animal I loved. She never failed me from weariness, carried me as fast as it ever became necessary to ride, and as easy as the rocking of a cradle, through many long and weary journeys, and under the protection of a kind Providence through dangers seen and unseen. And I cannot do less in giving this account of my journeyings than pay this affectionate and merited tribute to her memory."[20]

Dolly Spanker came into his possession at Bent's Fort, where he desperately needed to find another mule to replace one that had "given out." He discovered that mules were scarce, but he was fortunate to obtain one through the assistance of Marcellin St. Vrain, youngest brother of Céran St. Vrain, trader and partner of the Bent Brothers. Marcellin had found her "tricky and headstrong." Webb paid $20 for her, and "she proved just as he told me, very naughty, but very wise, easy riding, fleet of foot, and never tired. I became very much attached to her and crossed the plains several times with her and had no other riding mule."[21] After reaching Santa Fe, Dolly was sent with the rest of the herd to graze in a large pasture and was later declared missing, maybe stolen. Webb did not hear anything more about her. By chance, he happened to be in Santa Fe in 1850 and recognized Dolly standing in the Plaza in front of the St. Vrain store and was able to get her back. However, as a helpful gesture, Webb allowed a wagon master to borrow her for his return trip to Missouri. This time, Webb's loss was permanent; Dolly was supposedly stolen by the Pawnee somewhere along the Big Bend of the Arkansas River.

———

A growing number of Missouri farmers gained experience and skills in handling Mexican jacks and jennets and in producing larger mules by crossing the jacks with the robust horse mares, which settlers from Kentucky and other states brought with them. They also bred larger European jacks with horse mares and produced the famous Missouri mule. These mules began filling a burgeoning need for them on farms and plantations in Missouri and the South, across the country, and throughout the world. Mule breeding was a thriving industry by the 1850s, and by 1880, Missouri had become the leading mule-producing state in the country. According to the 1890 U.S. Census figures, the Missouri mule population had reached 245,273, the highest number among all mule-producing states.[22]

There were probably hundreds of stories about mules from the days of the Trail, a large number now forgotten. One story came to light in recent times about a mule that died when a steamboat sank. Mules, horses, and other animals taken to Missouri and the frontier by their owners were transported by steam packet, which was a steamboat traveling a regular route on a river and also carried passengers, mail, and freight. One of those was the *Arabia,* which ran the stretch on the lower Missouri River from St. Louis to the Port of Kansas (Kansas City). Its cargo was typical of the numerous items and materials carried over the Santa Fe Trail. It was one of about seven hundred steamboats navigating the Missouri since 1819. Of those, more than three hundred wrecks were strewn up and down the riverbed and in fields from St. Louis to South Dakota. Thirty-plus wrecks were clustered around the Kansas City area.

The *Arabia* sank in 1856 after hitting a snag hidden under the waterline of the Missouri. A May 30, 1897, article in the *Kansas City Times* reported river traffic to the Port of Kansas was heaviest from 1856 to 1858 because of the Santa Fe trade. The steamboat carried 130 passengers, none of whom were lost. During its excavation in 1988, a mule's skeleton was uncovered with bridle, saddle, and saddle roll still attached, and the reins were still tied to a post. Conflicting stories came to light. The owner of the mule, interviewed shortly after the sinking of the *Arabia,* said he tried to free his mule, but it was just too stubborn and would not leave the sinking boat. The skeleton of the mule tells a different story; the reins were tied as they had been for 132 years. Given the name Lawrence of the *Arabia,* the mule's skeleton, still wearing a bridle, is on display today at the Treasures of the Steamboat *Arabia* Museum in Kansas City.[23]

As agriculture, mining, and other industries turned to mechanization, mules were replaced by modern machinery, four-wheeled conveyances, and laborers. They began to disappear from farms and other places where their muscle power and endurance had been so critical to progress. Their numbers decreased sharply across the country. In recent decades, they have made a good comeback. Today, mules are again proving their broad range of abilities, many more than once thought they possessed. They are competing in every type of mule competition imaginable, including barrel racing, calf roping, steer stopping, cow cutting and penning, pari-mutuel racing, carriage driving, team chariot racing, and even high-level dressage. The lowly mule is being recognized as a champion!

Chapter Twelve

Burros (Donkeys) and Horses

The small Mexican donkeys called burros, larger donkeys, and all breeds of horses belong to the Equidae family. Since their domestication thousands of years ago, donkeys (*Equus asinus*) and horses (*Equus caballus*) have contributed immeasurably to the progress of humankind. Although they have been largely replaced in developed countries by the horseless carriage, the iron horse (train), and numerous other modern conveyances and labor-saving inventions, these animals, in particular donkeys, are still used in less developed regions of the world as beasts of burden and in agriculture. In the American West today, donkeys are used in a variety of ways, such as serving as pack animals in wilderness adventures and as guard animals for flocks of sheep. Horses continue to be important for recreation and sport, and as companion animals. They are also essential on working ranches of the American West.

The history of the domestication of the donkey is not known. There seems to be consensus that their domestication happened after that of cattle, goats, and sheep, and that their most probable ancestor is the Nubian subspecies of the African wild ass. The first donkeys to arrive in the New World came on a supply ship to Christopher Columbus on his second voyage in 1495. Four jacks (males) and two jennies (females) were included in an inventory of livestock delivered to Hispaniola, the Caribbean island now divided between the Dominican Republic and Haiti. They were bred with the Spanish horses to produce mules for expeditions and other purposes.[1]

Donkeys arrived in New Mexico with the first Spanish colonizing expedition led by Juan de Oñate in 1598. The Spanish word *burro* refers to a donkey. Burros are small, yet sturdy, strong, sure-footed, and patient. They have good eyesight and keen hearing. Their small hooves are very hard, and their favorite speed is slow and steady. They have an extraordinary ability to transport large, cumbersome loads on their backs with seeming ease. Thus, they have served as beasts of burden, pack animals, and transporters of people. Being very hardy animals requiring little care or attention, they have been expected to forage for their food to stay alive. Therefore, burros have long been known as the poor man's work horse.

According to George Champlin Sibley, the leader of the first government survey of the Santa Fe Trail beginning in 1825, burros were present on the Trail from its inception. During the same year, he and other treaty commissioners met with chiefs of the Osage at a place along the Trail. They named that place Council Grove, now an important stop in Kansas for modern-day Trail travelers. Sibley noted in his carefully written records that burros were among the animals brought back to Missouri by William Becknell and his small party of traders on their return from their inaugural trading expedition to Santa Fe in 1821. Sibley stated in a letter written in May 1825: "Becknell and his party returned home, having disposed of their merchandise to some advantage, the proceeds of which they brought home in specie [silver], mules, asses [the common name used for donkeys or burros], and Spanish coverlids or blankets."[2]

Trader Josiah Gregg described the burro in *Commerce of the Prairies:* "While I fully acknowledge the pretensions of the mule as an animal of general usefulness, I must not forget paying a passing tribute to the meek and unostentatious member of the brute family, the 'patient ass' or as it is familiarly called by the natives, *el burro.* This docile creature is emphatically the poor man's friend, being turned to an infinite variety of uses, and always submissive under the heaviest burdens. He is not only made to carry his master's grain, his fuel, his water, and his luggage, but also his wife and children. Frequently the whole family is stowed away together upon one diminutive donkey. In fact, the chief riding animal of the peasant is the *burro,* upon which saddle, bridle, or halter is seldom used. Seated astride his [the burro's] haunches instead of his back, the rider guides the docile beast with a bludgeon which he carries in his hand."[3] Gregg also mentioned that the burro, much like the goat, "sustains itself upon the mere rubbish that grows in the mountain passes, and on the most barren hills, where cows could not exist without being regularly fed."[4]

Travelers were often amazed at the burro's small size. Joseph Pratt Allyn, whose letters from the Trail appeared in the Hartford, Connecticut, newspaper, wrote while in camp along the Purgatoire River near Raton Pass on the Mountain Route in November 1863: "I saw some *burros* which are the little New Mexican asses, just about the size of a large dog. They are said to have extraordinary powers, endurance, and to live on anything. I have seen little donkeys in the east, but they were elephants beside these little *burros*."[5] Allyn had not yet seen anyone riding a burro and was puzzled about what a rider did with his legs if he rode one, because the burros looked so close to the ground.

Many other early accounts of travel over the Trail include mention of burros, especially in Trail travelers' descriptions of Santa Fé, the City of Holy Faith, where burros abounded in large numbers. They were a common sight on Santa Fe's Plaza, the western terminus of the Santa Fe Trail, and throughout the town and its outskirts. Strings of burros hauled firewood from wooded areas surrounding the town. It often took a full day to cut, load, and deliver the wood, which earned man and burro the sum of 25 cents.[6]

One who closely observed and described the burros roaming the dirt streets and alleys of Santa Fe was W. W. H. [William Watts Hart] Davis. He took the Trail to Santa Fe in 1853 to begin in his new position as United States attorney for the New Mexico Territory, and he lived in Santa Fe for two years. Davis kept a diary and later published *El Gringo; New Mexico and Her People* (1857). He saw innumerable burros during the time he spent in New Mexico: "We see before us an uncouth little animal . . . with a large load of wood strapped to his back, and urged along in a slow trot by a sharpened stick. His ears vouch for his relationship with the ass [donkey], and the name he bears is *burro*. In every particular he appears to be patience personified, and has been as highly favored with genuine ugliness as any species of the animal kingdom. They are small in stature, with a head wholly disproportioned to the body, with a pair of ears that should belong to a first class mule.

"But their virtues more than make up for their homeliness; they are the most useful animals in the country, and their services could not be dispensed with. They carry the marketing of the peasant to the towns to be sold, and bear their master home again; they carry the wife and children to church on Sundays, or whithersoever they desire to go; and if the country belle wishes to ride into Santa Fé on a shopping tour, she mounts her burro without saddle or bridle, and ambles off to town. They are capable of long fasting and much fatigue; they bear the most unkind treatment with the resignation of a martyr, and after a hard day's work will make a comfortable supper on thorn or cedar bushes, and their happiness is complete with a heap of ashes to roll in. They are made to serve innumerable useful purposes. . . . They are the universal hackney of the country people, who, when they mount the burro, instead of sitting on the back, sit astride the rump abaft [a nautical term meaning toward the stern or rear of a ship] the hips."[7]

Davis obviously had a soft spot in his heart for the little burro, because he added the following statement: "When New Mexico shall have become a state, the faithful burro should be engravened on the coat of arms as an emblem of all the cardinal virtues."[8] Although the burro was not engraved on the state's coat of arms, Davis would have been pleased to know that a statue of a burro laden with firewood now stands at the corner of Burro Alley and San Francisco Street, west of the Santa Fe Plaza and only a few blocks from the small gray granite "The End of the Trail" monument, which stands on the Plaza's southeast corner. In 2014, someone cut off the bottom half of the burro's tail. Being one of the most beloved statues of Santa Fe, local citizens and people in many other places were appalled at the news of such vandalism. A new tail was quickly put in place. The mindless, heartless perpetrator has yet to be found.

Narrow and a block long, Burro Alley runs between San Francisco and Palace Streets. Numerous burros were tied up or corralled here during Trail years. A large mural on the west side of Burro Alley depicts the bygone days when burros were kept there. On the west side of Burro Alley and facing Palace Street was the famous gambling establishment of Doña Gertrudis Barceló, also known as Doña Tules, a wealthy businesswoman of Santa Fe and the New Mexico Territory.

In addition to hauling firewood and transporting people, burros carried an incredible array of stuff on their backs. A rare book of photographs of burros taken by a number of frontier photographers was compiled by Richard Rudisill and Marcus Zafarano and published in 1979.[9] They explained in their foreword that early photographers found burros a profitable subject.

In the last decades of the Trail, photographs of burros were purchased by Trail travelers and tourists from the States. In the 150 black-and-white photographs ranging in date from 1866 to the 1930s, burros are shown carrying firewood and such diverse loads as heaps of hay and corn husks, kegs of the area's favorite whiskey called Taos Lightning, gramophones, and an ore car used in the mines. From the mountains, they dragged *vigas,* long wood beams used in the construction of buildings. They performed a myriad of laborious tasks and were heavily relied upon in mining operations until the mining boom faded and many mines closed.

Among the photographs of burros in Rudisill and Zafarano's book is one taken by Charles Fletcher Lummis, adventurer, author, and journalist, who survived his five-month "tramp across the continent" a few years after the Trail was replaced by the railroad. Lummis began his walk in 1884, starting out from Chillicothe, Ohio, and headed to Los Angeles to assume the position of city editor of the *Los Angeles City Times.* He began in his new position in early 1885. The photo, a close-up of a burro's head, was taken by Lummis in Santa Fe. According to Marc Simmons, "Lummis declared that, for its size, no town on earth had more burros than Santa Fe."[10] Other photos in Rudisill and Zafarano's book show what Burro Alley looked like in those days—a far cry (or better, a far bray) from how it looks today.

Besides noting the large number of burros in and around Santa Fe, Trail travelers and others sometimes wrote about their bray: the ear-shattering, deafening noise made by burros and other donkeys, especially when they woke up hungry every morning. Who needed a rooster to greet the dawn and awaken the whole town? Once they were able to roam about to find something to eat, their braying quieted down, although not entirely. Seldom were they ever fed by anyone. Being able to subsist on almost any vegetation, burros were inexpensive. The going price for one in Gregg's day was $1, and by the end of the 1800s, the price had risen only to $5.[11]

In the 1950s, when burros had almost disappeared from sight, Marc Simmons bought a pack burro for $40 at a sale barn in Albuquerque and named him Taco. He led Taco through the backcountry of northern New Mexico, reliving what he calls the last Age of Donkey Power. Simmons has many good memories of their trips: "I considered my experiences with Taco an authentic way to get in touch with New Mexico history. During the many hours spent alone with him on old trails, I learned much about the temperament and habits of burros. The experience increased my respect for these smart little animals."[12]

Although patient and generally cooperative, Taco could be stubborn and unwilling to do certain things, such as stepping into a creek or fording a stream. He would stop at the edge of the water, brace his little front hooves, stiffen his legs, and refuse to budge. With some pushing, pulling, and coaxing, he would finally loosen up and take a step. Once he got his feet wet, Taco went across without any trouble. Simmons also found the little burro amusing. For instance, Taco carried Simmons's food supply, which usually included a few grapefruits, one of the burro's favorite treats. After eating the grapefruit wedges, he gave the juicy half-shell to the eager burro. Taco always took the half by biting it on the edge, and it would invariably flip up and over his nose. Taco stood there stoically, wearing the grapefruit shell on his nose. This never failed to make Simmons laugh. He also mentioned that burros have an amusing way of scratching an ear with a hind hoof.[13]

A number of breeders and owners think burros have an intelligence superior to that of the horse. In his excellent book *The Burro,* Frank Brookshier told about ways in which they differ from horses: "Unlike the horse, the burro never cuts himself on barbed wire by running into it as a horse will do. Even when frightened, the burro does not lose his head to the extent that he dashes into such a 'barbarous' obstacle. If one of his legs does become entangled in wire, he will not fight or attempt to extricate himself like the horse does. Instead, he will remain patient and wait for his master to come to his rescue."[14] Burros are also very watchful, making them good guards, particularly for sheep herds. Even though small in size, they are not fearful of a dog or a coyote and will run off such intruders.

As the age of the burro was ending, large numbers were abandoned to fend for themselves across the land. Some of those burros joined the free-roaming feral bands living across the western deserts and on public lands.[15] The Wild Free-Roaming Horses and Burros Act of 1971 gave burros a legal right to live on public lands without harrassment. They were also included in the Bureau of Land Management's Wild Horse and Burro Adoption Program, which was established in 1973 and has found good homes for many of these wild animals.

In more recent years, at least one donkey has walked along the entire old Santa Fe Trail. Nebraskan Les Vilda walked over the Trail in 1984 with three other walkers and his faithful pack donkey named Joker. The donkey packed food, water, and a variety of items that included a bed roll, cooking pot, camera, change of clothes, and a brush, hoof pick, and watering bucket for Joker. The group walked from Fort Osage, east of Independence, Missouri, to Santa Fe, covering 980 miles in ninety-one days, which consisted of eighty walking days and eleven rest days. There was rain only on two days of the entire trip over the Mountain Route.

Vilda returned to the Trail in 1987 with Joker as his sole companion. Originally planning a round trip over the Mountain Route, he and Joker left from New Franklin, Missouri, on April 20. By the time they covered one hundred miles, Joker had become lame and had to be taken back to Nebraska in a trailer to heal. Vilda brought back to Kansas City a small covered wagon about four feet wide and six feet long, and bought a horse named Zulu. Vilda recalled the experience in *Wagon Tracks,* the quarterly of the Santa Fe Trail Association: "The horse was broke, but I was not. For several days, I received pointers and lessons on how to harness, hitch, and drive. This was done on the streets of Kansas City, and I probably learned quicker than most people, because of the fear of getting killed had I done something wrong. It took close to a week before I felt confident enough to set out on my own on the Trail."[16]

Vilda found that wagon travel had both advantages and disadvantages: "One of the better points was that I could carry some luxuries that I couldn't before." Those luxuries included more food and a mattress (straw-filled), which allowed him to sleep in the wagon, rather than on the ground, and to have cover when it rained at night. He kept his rolled up mattress and blankets under the wagon seat during the day. After surviving a terrible storm outside of Tecolote, New Mexico, and numerous careless, inconsiderate drivers on the roads, Vilda and Zulu reached Santa Fe on September 10, 1987, having spent ninety days along the Trail. Those days consisted of sixty-three traveling days and twenty-seven rest days. Rain fell on thirty-two of those days. Vilda recalled there were "very few 'Ho-Hum' days on the Trail," because he "could always count on either Zulu or Mother Nature to come up with something to make the day exciting."[17] He decided to not make a return trip, as he had first planned, because of the dangerous traffic on the highways he had to travel.

———

Many travelers rode their horses along the Santa Fe Trail. Stanley Vestal wrote about the importance of the horse on the Trail and the Plains of the Far West: "There man and horse became one. The animal was no longer a tame creature of stall and barnyard, but the companion of his wandering master, a friend in need, indispensable in war or peace, something precious that a man must risk his neck in guarding, something irresistible, worth a man's neck to steal or capture. Many a man was hanged or shot for horse-stealing on the Plains. . . . Many a man broke his neck in a tumble from the back of his horse. The saddle was, of all places, the most dangerous—yet also the happiest."[18]

Horses were used by William Becknell and his companions on their first trading expedition to Santa Fe in 1821. They were used again in 1822 and 1823 and also to pull the first wagons over the Santa Fe Trail. However, horses proved inadequate to meet the rigors of pulling large and heavy wagons on the long, rugged routes. For example, George Sibley and the first team surveying the Trail experienced numerous problems with their horses tiring, becoming sick, and even giving out, unable to continue pulling their wagons. After much consideration about how to proceed, a decision was made to procure mules from Taos. The mules carried the baggage removed from the wagons, and the empty wagons were hauled by horses over the mountains. Only in that way was Sibley's survey team able to complete their way to Taos, thereby saving their horses and wagons, and also proving the existence of a wagon route over the mountains to Taos.[19] By the beginning of the 1830s, horses had been replaced by mules and oxen for pulling heavy freight wagons.

There were, of course, other important uses for horses, in particular hunting, riding, and the chase. Horses pulled light wagons, carriages, and stagecoaches, although mules were often preferred for these and other modes of transportation. Most men in a caravan had a horse with them for riding and hunting, either bringing their own horse or purchasing one in Kansas before their caravan was ready to journey westward. For example, W. B.

Napton, eighteen years old and yearning for an exciting adventure, left his home in Saline County, Missouri, and headed to Kansas in 1857. He arrived in Westport, where he and other travelers had to wait for two weeks for the arrival of freight at Kansas City. There were no tents in his camp situated three miles southwest of Westport, so he slept under "the broad canopy of Heaven." During their wait, Napton acquired a horse: "I was fortunate in purchasing a first-rate 'buffalo horse' that had been brought across the plains the previous year. He proved his excellence."[20]

Some women had their favorite horse with them, also. One was a young Army wife, Eveline M. Alexander, an expert rider, who was also able to shoot a pistol with accuracy. She married Union Army officer Andrew Alexander in 1864 and left her home in Utica, New York, in early 1866 to join him in Arkansas at Fort Smith, where she arrived in May. Her beloved horse, Zaidee, also called Zay, and their dogs were already there with Andrew. Eveline rode Zaidee frequently during the long, tedious journey from Arkansas to forts in Colorado and New Mexico. She also accompanied her husband on hunts for quail, deer, and other game. Sometimes she rode in the ambulance, which belonged to her husband's cavalry regiment. An ambulance was a heavy spring wagon with a canvas top and leather-upholstered seats. It was primarily used to carry the wounded and sick, but also served as a passenger wagon for the officers and their families. She disliked the jolting she received riding in it and to her relief she was not in it when it later broke down. Eveline preferred riding Zaidee.

On one occasion, Eveline was invited to review the regiment: "Colonel Howe insisted I should ride by his side during the review and wanted me to receive the salute as reviewing officer, but the last compliment I declined. I rode Zaidee . . . and wore my grey riding habit and black velvet hat. Zaidee behaved beautifully. After the regiment had passed in review before the colonel, we rode along the line and, as the ground was rough and full of ditches, Zay distinguished herself. The first ditch we came to was rather wide, but she took it like a bird and behaved throughout with the most perfect decorum. Zaidee was very much complimented on her behavior and appearance."[21]

When approaching Fort Garland in Colorado in October 1866, Eveline was met by Andrew, who was riding in General Kit Carson's carriage. The couple were invited to have supper with the general and spent the night with him. She also dined with "old Kit Carson" the following day; Andrew was unable to join them because he was sick in bed with a bad cold. The next day, October 13, Eveline wrote in her diary: "This morning I went out with Kit Carson to visit the Indian encampment about five miles out. I rode Zay and she never appeared fitter. General Carson declared she was the finest animal he had ever seen." [22] Eveline was always proud of her beautiful Zaidee.

Many men were very proud of their horses, too. They paid homage to the horses for their valor and getting them through rough, dangerous places and times. For instance, Joseph Pratt Allyn paid homage to his stable-bred horse, Swindle, after his journey over the Trail was completed. He was heading to other points in New Mexico Territory and wrote in a letter dated December 1863 from Fort Wingate in northwestern New Mexico: "Just before leaving Albuquerque, I exchanged the horse I had ridden every step of the way from the States. Poor Swindle deserves a passing notice, noble, clumsy, affectionate animal that he was. He received his name to commemorate an attempt on the part of a quartermaster to prevent my using him. He was large, a hard trotter, clumsy at a leap, glorious in a run, possessed of wonderful powers of endurance, indeed superior to any other in the outfit. I never knew him to stumble in the thousand miles I rode him, and together we have been over mountain paths, when a single misstep was swift destruction to horse and rider.

"At Fort Union I was offered the pick of the corral for him, but declined to exchange him, believing that the rest at Santa Fe would put him in condition to go through. Two long weeks he shivered in the snow, then on half rations of wheat, and at the end was worse than at the beginning. It could not be helped; there is scarcely any forage in New Mexico, and that was the real reason General Carleton sent back most of our cavalry escort. . . . I left Swindle at Los Pinos, below Albuquerque, where there is a government depot, good grass, and fine corrals. The poor brute looked at me as though he wanted to try to go, and I plead guilty to brushing away a tear as I parted forever with the faithful creature that had shared the dangers and privations of a journey I am not likely to forget." [23]

Many horses were lost along the Trail in stampedes, run off by renegades, or ran away on their own. Large numbers, like the other animals in the wagon trains, perished in freezing weather and from disease, injuries, hunger, thirst, and exhaustion. Marion Sloan Russell told about the loss of the herd of horses being taken to Fort Union in 1852 during her first trip over the Trail, when she was only seven years old. The large caravan in which she, her mother, and brother were traveling was camped overnight at Pawnee Rock in Kansas. During the night, Indian war whoops broke the silence, and the camp became tumultuous. Something struck the top of their tent, which collapsed upon them. She later recalled: "When morning broke our horse herd was gone, a herd of two hundred army horses. The Indians had stampeded them and driven them off in the darkness. Captain Aubry [François X. Aubry, freighter and record-breaking long-distance rider] would not go on to Fort Union without the assignment of horses, so we went into camp at Pawnee Rock and waited two weeks while the out-riders went back to Fort Leavenworth to buy up more horses. Each day as we waited we were visited by roving bands of Indians."[24] When the herd arrived, the caravan proceeded and safely reached Fort Atkinson on the Arkansas River. From there, they headed without incident to Fort Union, which had been established in New Mexico in 1851.

Sometimes a lost or missing horse was found, although such an occurrence was considered to be like finding a needle in a haystack. This happened in 1864 when James Francis Riley's wagons were being loaded with government supplies at Leavenworth, Kansas. He thought a horse hitched to a rack looked familiar and went closer to get a better look. Sure enough, it was Old Bill, the horse his family had brought from Ohio. Old Bill had been stolen, taken to Missouri, and sold there to a man who lived on the road to the town of Lawrence, Kansas. Riley recounted what happened next: "At once we got out the papers to replevin Old Bill. The man conceded our right to him. We let the man take Bill home and keep him until called for. So, as we passed by his place we took Old Bill along with us. He was an old standby, being over twenty years old and one of the first horses I ever worked in the spring of 1854 in Ohio. I also drove him to Kansas in 1857, so you see he was like one of the family."[25] Old Bill was with them until he died of old age at twenty-six or twenty-seven.

The number of domestic horses on the Trail is not known; the total number of domestic horses, feral horses, and wild mustangs had to have been in the millions. The U.S. Department of Agriculture reported a historical domestic horse population figure for 1867 of eight million. One hundred years later, in the 1960s, that figure stood at three million.[26] Horses were indispensable in so many ways and for so many reasons. They made their mark on the Santa Fe Trail and every direction beyond.

Chapter Thirteen

Dogs

Humanity's best friend, the dog (*Canis familiaris*) of the Canidae family, is the oldest of all domesticated animals. The prehistory of the dog, going back thousands of years, is shrouded in the mists of time. Only in recent years, especially with the advent of DNA research, have we begun to learn about the transformation of the wolf to dogs, now numbering more than four hundred breeds. There is increased interest in the history of their origins and in the human-dog relationship. This is proper and fitting in light of humans' close interactions with dogs. More than thirty-five years ago, historian Marc Simmons pointed out that the contributions of the dog to the history of the American West were being overlooked and neglected.[1]

There were numerous dogs on and near the Santa Fe Trail in the 1800s. Their names were short, and among the more popular ones were Bob, Boots, Duke, Jack, Lee, Toots, Watch, and Zeus. They were primarily large, muscular, strong dogs and, for the most part, alert, intelligent, and faithful. There were a variety of pure breeds, in particular bulldogs, greyhounds, Newfoundlands, pointers, setters, and Saint Bernards. These are the breeds most frequently mentioned in diaries, journals, and narratives of Trail travelers. They were useful as hunters, trackers, guards, protectors, and companions. There were, of course, smaller breeds, mixed breeds, and what has been referred to as the standard mutt.

The dogs of the Indian tribes of the plains, having inhabited this vast land for thousands of years, ranged in size from small to large and were invaluable as watch dogs, hunters, and draft animals carrying packs on their backs or dragging the *travois,* a platform mounted on two long poles for carrying goods or small children. The long hair of some dogs was used for weaving, until the arrival of sheep brought by the Spaniards. The dogs also performed a variety of tasks, such as turning spits for cooking meat, pest control, and

garbage patrol. Dogs also served as a food supply, in particular for special feasts, among certain tribes, and as the number of wild animals, especially buffalo, declined. In freezing winters, they were useful as foot and bed warmers, which a number of Santa Fe Trail travelers also appreciated. The Indians also kept dogs as companions or pets.

Sometimes dogs' lives were short. Many died along the Trail because the trip was as hazardous for them as it was for humans. Death came in a number of ways. Some were biological: overwhelming thirst and dehydration, starvation, and disease. Others were circumstantial: being trampled in stampedes, crushed under wagon wheels and the hooves of domesticated animals, killed by wild animals, or shot and killed accidentally—or not so accidentally in skirmishes and battles. A number collapsed and died from heat exhaustion following a long chase after buffalo or other animals in hot weather.

Lydia Spencer Lane, among the number of military wives who traveled over the Trail, wrote about bringing Lee, her Saint Bernard puppy, on one of her trips and losing him in a fatal accident: "He was growing fast and, of course with his cunning ways, was a great pet with everybody. He was put into our wagon on leaving camp, where there was someone to look after him. But one day the watchman went to sleep, and our poor puppy crawled out of the wagon, fell under the wheels, and was killed instantly. There was great sorrow and indignation in the camp when it was known Lee was dead, and the soldiers who pitched our tents would not allow the man who had charge of him to come about the place. I cried all day for my puppy, and never would have another."[2]

Dogs were often not welcomed by wagon masters, because a dog might bite or frighten the mules, oxen, or horses and cause a stampede. Also, dogs could get underfoot and cause injury to the people in a caravan. Men would quarrel and fight when a dog belonging to one of them bit the other man or one of his animals. Veteran wagon master, Tom C. Cranmer, stated in his *Rules and Regulations by Which to Conduct Wagon Trains* (1866): "Never allow a dog in the train."[3] Such sound advice was frequently overlooked. Many on the Trail would not consider being without their canine companion.

———

Dogs would sometimes become lost along the way. If they were good trackers, using their keen sense of smell, they might find the wagon train. Often, the lost dogs returned in pitiful condition. One traveler who wrote about a lost dog was popular author Washington Irving, who traveled with an aristocratic hunting party and afterward published *A Tour on the Prairies* (1835): "Pursuing our journey, we were met by a forlorn, half-famished dog, who came rambling along the Trail, with inflamed eyes and bewildered look. Though nearly trampled upon by the foremost rangers [riflemen who rode their horses ahead of a caravan], he took notice of no one, but rambled heedlessly among the horses. The cry 'mad dog!' was immediately raised and one of the rangers leveled his rifle, but was stayed by the ever-ready humanity of [our leader], who exclaimed, 'He is blind! It is the dog of some poor Indian, following his master by the scent. It would be a shame to kill so faithful an animal. . . . ' The dog blundered blindly through the cavalcade unhurt, and keeping his nose to the ground, continued his course along the trail."[4]

Writer and actor Matt Field joined a wagon train in 1839 as an adventure with a group of friends. He mentioned in his journal that there were two dogs in their caravan. Interested in learning some Spanish during his trip, Matt sometimes used it in his writing and for dating his daily journal entries, two of which mentioned a lost dog: "*Sabado* [Saturday] 5th—Travelled along the Semirone [Cimarron River] from 6 a.m. till *Trece y Media* [3:30 p.m.]. Saw a Stray Dog"; and "*Domingo* [Sunday] 6th—Long, flat, tedious prairies, grass crop[p]ed short, ground hard, strong wind. Found fresh horse tracks—Stray dog still following us."[5]

In a series of articles about his Trail experiences for the New Orleans *Picayune*, Matt wrote one titled "The Lost Dog." In this article, he told about venturing out from the camp on the banks of the Arkansas to get a look at a wolf that appeared nearby. His party soon discovered that the wolf was a dog, "a nearly starved, timid, domestic creature, which had been lost probably by some solitary trapper or wandering Indian hunter." He also thought the dog might have been lost from a passing caravan. Although he and his companions tried to coax the forlorn creature nearer to give it some food, the lost dog was full of fear and ran from them. Field recalled: "It still turned to gaze at us, and rather *slunk* than ran."[6]

The next morning a night guard reported that a wolf had approached him within twenty feet, but he could not shoot his gun because it would alarm the camp. The following night, all were awakened by the fierce barking of the caravan's dogs, and every man was instantly awake with rifle ready. The cause of alarm was "the same poor starving animal" following them. The forlorn dog had crept into camp in search of food. Field's party left bones and scraps of meat behind whenever they struck camp.

For five days the lost dog followed the caravan, always crouching at a distance when it stopped. When the dog seemed to want to come nearer, the two dogs with the caravan chased it away. Field recounted: "This poor dog had been wandering about the prairie evidently a long time, for when it was at last brought into camp we could perceive it was dwindled almost to a skeleton, and its extreme shyness towards us sufficiently proved that it had endured much. Misery and luxury are equally potent in making cowards, and the rule applies to dogs as well as men."[7]

Field concluded his article: "The poor animal we had picked up in the wilderness, followed us through the remainder of our travel, till we reached the first log house that appeared among the far western settlements of Missouri. Here we gave him to a farmer, and as we sat beneath the hospitable shelter . . . feasting upon a luxurious banquet of cornbread, fried bacon, and rich milk unmingled with water, we told the history and adventures, and excited the good farmer's sympathy for our poor desert foundling, the lost dog."[8] If anyone could convince the farmer to adopt the dog, it surely would have been kindhearted, eloquent Matt Field.

———

The best-known dog in the history of the Santa Fe Trail was probably Susan Shelby Magoffin's greyhound, Ring (short for Ringling). Susan first mentioned Ring near the beginning of her diary when she described the wagon train belonging to her husband, trader Samuel Magoffin: "We now numbered, of ourselves only, quite a force. Fourteen big waggons with six yoke each, one baggage waggon with two yoke, one dearborn with two mules (this concern carries my maid), our own carriage with two more mules, two men on mules driving loose stock. . . . With Mr. Hall, the superintendent of the waggons, together with his mule, we number twenty men, three are our tent servants (Mexicans). Also, Jane, my attendant, two horses, nine mules, some two hundred oxen, and last though not least our dog Ring. A gray hound he is of noble descent; he is white with light brown spots, a nice watch for our tent door."[9] There were forty-five wagons in the entire caravan, which left Council Grove in present-day Kansas on June 21, 1846, and arrived in Santa Fe on August 31, 1846, only two weeks after General Stephen Watts Kearny had peacefully occupied the city.

Susan wrote while in camp: "Last night I had a wolfish kind of a serenade! May Pan preserve me from the likes tonight. . . . Ring, my dear, good dog!, was lying under my side of the bed, which was next to wolves. The instant they came up, he had been listening, he flew out with a fierce bark, and drove them away. I felt like caressing him for his kindness, but I had another business to attend to just then. Rid[d]ed of our pest [the wolves], I was destined to suffer from another." She was referring to "the winged pesti-lence" of the Trail, known as "musquitoes." While fighting off the mosqui-toes, Susan wished "my faithful Ring would not sleep so soundly. Just then, as if he had heard my thoughts and was anxious to prove to me that I was too hasty in my decision as to his vigilence, he gave one spring from his hiding place, and in a twinkling had driven them off entirely. As lonely as I was, I laughed outright. Sleep had entirely deserted me, so I '*kept watch*' till daylight. All the morning I have been nod[d]ing."[10]

While walking in a thicket of grapevines along a stream near Council Grove, Susan suddenly became very fearful, but Ring came to the rescue: "Ring, my faithful Ring, came by me just then and I commenced patting his head which made him lie at my feet and I felt safe with this trusty sol-dier near me."[11] When the caravan was "out on the prairie with no wood and little water" one night, the chilly damp air prompted her to look for her bed cover. "After searching in the dark, I found that Mr. Ringling had very gallantly made his bed on it. I hoisted him from this berth though, and with my burthen [burden] crawled back to my own, to '*make the best of it*.'"[12]

When they were in buffalo country, Susan wrote at Big Coon Creek: "Passed a great many buffalo (some thousands), they crossed the road frequently within two or three hundred yards. . . . Ring had his own fun chasing them." When Ring got near a buffalo, "he [the buffalo] would whirl around and commence pawing the earth with not a very friendly feeling for his deli-cately formed persuer [*sic*]."[13] After reaching "the entrance of what is called the 'Raton,' a difficult pass of fifteen miles through the Mountains," she noted in her diary: "Our tent is stretched on the top of a high hill, at the foot and on the sides of which I have been rambling accompanied by our faithful Ring, who all the while kept strict watch for *Indians, bear, panther, wolves* &c. [etc.], and would not even leave my side as if conscious I had no other protector at hand."[14]

One more "ramble" with Ring on the mountain was recorded. It was the last mention of faithful Ring in Susan's diary. One wonders what happened to the dog. This is a common occurrence in other diaries and memoirs. A dog may be written about from time to time, but then there is no further comment. Did those canine companions survive traveling the Trail? Did something happen, and it was not recorded? Was there so much sorrow and grief that the death of a beloved companion was not mentioned?

Katie Bowen wrote, in less glowing terms than Susan Magoffin's, about the dog traveling with her and her husband, Captain Isaac Bowen, from Fort Leavenworth to Fort Union in 1851. Katie commented about the dog in a letter sent from Fort Leavenworth on April 15, 1851, before their departure: "Isaac's dog is a bouncer. We call him Bruno. He will be a fierce protector."[15] In early July, she included in a letter: "We keep our big dog chained to the waggon in front of our tent at night, but he must have slept wondrous sound last night for a wolf went to the fire and ate what scraps were left in the pots, rattled among the tin pans so that the teamsters were disturbed, and then put out for the herd."[16] Later that month, she complained: "Our great worthless dog gave out and we had to take him in, quite to my disgust, for my love of dogs is not great."[17] Poor Bruno may have "given out" because July was one of the hottest months on the Trail. He could have been suffering from heat exhaustion and dehydration, which severely affected animals, as well as people.

Many hunters and sportsmen traveled over the Trail and brought along their favored breed for hunting, tracking, and retrieving. They had strong opinions about the best dog for hunting. One of those men was Colonel Richard Irving Dodge, an avid hunter and confirmed admirer of pointers, a breed favored for its strength, speed, smartness, keen sense of smell, and short hair. He wrote in his popular book *The Plains of the Great West and Their Inhabitants* (1877): "A well trained dog is most invaluable to the sportsman, for whatever his skill as a marksman or trailer, he will lose more or less game unless he has the assistance of man's best friend."[18]

Dodge found pointers especially useful for tracking deer and other game. He rejected setters because they have long hair and could not stand the heat and dryness of the Plains. He rejected greyhounds and bloodhounds because they were "stupid" and curs and foxhounds because they were "difficult to train." He did admit, however, that even his beloved pointer, "like a dimwitted enthusiastic junior officer, was not always totally under control."[19]

Richard I. Dodge helped found Dodge City, Kansas, in 1872. The city was named for Fort Dodge, which had been named for its founder, General Grenville Dodge, no relation to Richard. After traveling to many forts and seeing numbers of dogs, Col. Dodge stated: "Soldiers were proverbially fond of dogs and, as a result, most forts resembled breeding kennels featuring countless dogs of seemingly limitless variety."[20] He also thought the majority of those dogs consisted of curs, lacking "the snap and obedience" demanded by him and other military officers.

Mountain men and trappers had a dog with them wherever they went. William Bent, younger brother of Santa Fe Trail trader Charles Bent, was in Santa Fe in August 1830. He headed out with three other men to trap beaver. Each man rode a horse, and they took along nine pack mules, twenty-four traps, a Mexican roustabout, and a huge dog named Lolo. On their return, the successful trapping trip suddenly turned into a dire situation that ended with the five men defending themselves against a large number of Indians. After thirty minutes of ferocious fighting, the Indians lost interest and left with their injured and dead men. Lolo, William's dog, contributed by whipping two of the Indians' dogs. The trappers and Lolo rapidly headed for the nearest Spanish settlement, poorer than they had left—except in experience.[21]

Among the trappers who spent time at Bent's Fort was William Bransford, known for his red whiskers and unusual ways. He became a friend and hunting companion of William Bent and Tom Boggs, a member of the family for which Boggsville in southeastern Colorado was named, and a grandson of Daniel Boone. A story was often told at the fort about Bransford when he was still a greenhorn. Once when hunting buffalo, Bransford tried to kill a big bull by dismounting his horse directly in front of the animal and shooting it three times in the forehead. At that time, he did not know a buffalo could not be killed that way, and as the bull charged closer, Bransford had to scramble, while the party "hoorawed" (cheered).[22] However, he later proved he had some amazing skills with animals.

David Lavender told in his acclaimed chronicle, *Bent's Fort* (1954), how Bransford happened to acquire a dog at the famous fort: "A huge white dog, as big as a Newfoundland, appeared one evening with the wolves that always prowled outside the walls for refuse. The men lured the dog inside the entry with a bait of raw meat and slammed the gate shut behind it. Promptly the beast went berserk and chased every person in the fort up on the roof. From that vantage point they managed to lasso the dog, put it in iron chains, and locked it in the fort's bastion. For two days it stayed there, howling and leaping bare-fanged at anyone who approached."[23] Finally, on the third day,

Bransford declared in no uncertain terms that enough was enough, walked up to the ferocious dog, and patted its head. From that time on, the dog was Bransford's, but no one else dared to touch it, and the traders ceased their joking and wisecracks about Bransford's earlier "buffalo episode."

Bransford became a legend in his own time for his way with the huge dog, his constant companion. Among his human companions was adventurous seventeen-year-old Lewis Garrard, who in 1846 and 1847 traveled the Santa Fe Trail, into the Rocky Mountains, and to Taos. He, like Bransford, was enamored with the joys of a hunter's life. During a hunting trip up Timpas Creek, close to the Mountain Route in southeastern Colorado, a snowstorm stopped the party from going on, and for two nights they had no firewood. To stay warm during those freezing nights, Garrard had to sleep with Bransford, but Bransford first had to convince his big, ferocious dog that a guest was acceptable.

Garrard wrote about the same trip with Bransford and his dog in *Wahto-yah and the Taos Trail*. He described the dog as being "of the Mexican-shepherd breed, somewhat larger than the Newfoundland, with white, coarse, and long hair." Neither man mentioned the dog's name, if it had one. Garrard also described how the dog had been baited with meat and trapped at Bent's Fort: "The infuriated dog, far more strong and savage than the largest wolf, turned on his assailants, who safely stood on the roof . . . with lassos to noose him as he ran from one side to the other. The dog bit several lassos in two. He was fastened with a heavy iron chain in the bastion. On the third morning, Bransford walked up without show of fear, patted him on the head." Regarding those freezing nights without firewood, Garrard wrote: "Bransford and I occupied the same bedding; and, if I wanted to sleep first, he had to accompany me until safe between the blankets, for the faithful guardian always asserted his right to coil himself on Bransford's property and growl at those who approached. His warm body was quite an acquisition to the comfort of our feet."[24]

Garrard's description of the dog as "of Mexican-shepherd breed," may have referred to the New Mexican sheepdog that had a bloodline reaching back to Spain. According to Marc Simmons, when the Spanish colonists brought their dogs, among them very large greyhounds and mastiffs, to the

Rio Grande Valley, they crossed theirs with the Indians' large dogs. The new mix was larger and tougher than better known shepherd breeds. A very young pup was taken from its mother, suckled by a ewe (a mother sheep) and grew up with the flock of sheep, developing a strong sense of attachment and responsibility for the flock. Those dogs effectively kept wolves and coyotes at bay, and when trained could control hundreds of sheep, which became a mainstay of the frontier economy in New Mexico. This breed is generally now considered extinct.[25]

Numerous dogs were seen by Garrard when he stayed in Indian villages. He wrote about one of his visits with the Cheyennes: "In this village were more than a hundred dogs—from the large half-wolf down to the smallest specimen. Often, during the night they broke forth in a prolonged howl, with the accompanying music of hundreds of prowling wolves making a most dissonant, unearthly noise."[26] He added: "Frequently when we executed a song in our very best manner, the village dogs chimed in with their original and touching music, forcing us to acknowledge ourselves beaten, in fair fight, and to withdraw leaving them undisputed masters of the field."[27] As Garrard noted, the Indians had dogs in a variety of sizes, shapes, and colors. Trained by the women, the larger dogs transported the Indians' possessions on the *travois,* and medium-sized dogs carried goods in packs on their backs until the horse arrived and eventually replaced the dogs as transporters.

Dogs were used for food by some tribes, while others refrained entirely from eating dog meat. European American and European travelers dined on dog meat and wrote about the experience, some stating that the meat was delicious and nourishing. Among those was Trail merchant Theodore Weichselbaum, who traded with the Cheyennes, Arapahoes, and Kiowas for buffalo robes and antelope (pronghorn) skins. In his 1908 narrative of his adventures along the Trail in 1864, Weichselbaum told about being invited by the Cheyennes to trade with them in their camp. He and his partner spent four weeks at the camp: "The Indians treated us well. Their camp was south of the Arkansas—a great big camp. . . . They furnished us a lodge to live in." The partners dined on hearty soup, roasted buffalo, a sausage made of buffalo meat and red berries, and a dish of "little dogs roasted." Those little dogs, according to Weichselbaum, were raised for that purpose and "just as nice and fat as could be."[28] Garrard described the cooked meat as being rather uninviting to the eye, but delicate, sweet, and reminiscent of cold roast pork.

———

In *My Life on the Frontier, 1864–1882* (1935), Miguel Antonio Otero, Jr., the first Hispano governor of the New Mexico Territory, told of his love for animals, in particular dogs and horses. His favorite diversion was hunting, especially antelope (pronghorn), for which he kept a pack of sixteen grey-hounds. He lived in the area of Granada in southeastern Colorado, actually at New Granada, which was the terminus of the Santa Fe Railroad at that time and where the prominent mercantile firm of Otero, Sellar & Company was built to supply Fort Union. The Granada to Fort Union Military Road was used by freighters headed for the fort and Santa Fe.

Among the numerous dogs Otero owned during that time was Duke, a fine English bulldog, who became a close comrade of Kiowa, Otero's beloved blue roan pony, once a wild mustang and later a champion racehorse. Duke stayed with Kiowa day and night and would not let anyone other than family members approach the pony. During spring and summer months, Kiowa was usually picketed with about 125 feet of rope near Otero's house. The iron picket pin came loose one day, and Kiowa joined a herd of horses passing very close to his home, "thus exposing himself to every chance in the world of being conducted into the Indian territory, whence the herd was bound. Not until late in the afternoon did any of us notice Kiowa's absence, and when we did, we discovered that Duke was likewise absent. Not a trace of either could we find during the next few days, although we made a diligent search."[29]

Otero was about to come to the conclusion that he had lost Kiowa and Duke forever. It happened, however, that some freighters returning from the Canadian River passed by. In response to Otero's queries, they reported seeing the herd and had noticed a horse following it at some distance and dragging a long rope with a bulldog pulling on the end. It was a stroke of luck for Otero, for the road from Granada to the Canadian River was seldom traveled. Immediately, he began preparations to hunt for the two, when "in walked Duke leading Kiowa with the rope! Duke was tired and haggard, clearly showing that he had suffered terribly on the trip, which had taken ten days. I doubt very much whether he had had anything substantial to eat since he left home. The pony, on the other hand, looked as fine as usual, for he had found plenty of grass and water."[30]

Throughout the rest of his life, Otero believed there had never been a better example of an animal's love and affection—and determination: "There is no doubt in my mind that when those freighters saw Duke tugging at Kiowa's rope, he was trying to turn the pony in the direction of Granada, and I am sure he never gave up until he had accomplished his purpose. When he had finally induced Kiowa to accept his views, he took the end of the rope and led Kiowa back to Granada. I learned later from reliable sources that dog and pony had followed the herd of horses for more than 150 miles before Duke could change Kiowa's mind." Otero stated: "I am greatly beholden, for it kept me from losing Kiowa."[31]

Duke regained his usual good condition, only to be poisoned a few weeks later by an inhumane brute. Otero remembered with sorrow: "We did all in our power to save him, but the poison had already done its work." He felt that Duke understood they were trying to help him, and would never forget "his pitiful and pleading look and the way his eyes said thanks for all we were doing."[32] Duke was buried in the yard close to Kiowa's stable, where the two friends had enjoyed each other's company. Kiowa lived to the age of thirty, dying of old age.

Dogs have, indeed, a long and honorable history. They helped and served their human companions in countless ways on the Santa Fe Trail, across the West, and around the world. Like Governor Otero, people throughout history and today are beholden to dogs for their faithful service, courage, and companionship.

———

A Brief Note about Cats

After reading about dogs, one may wonder if cats (domestic felines) traveled on the Santa Fe Trail. Most cats would never put up with such a long, arduous, uncomfortable trip. They would head lickety-split for the nearest settlement, preferably one with a milk cow. However, some cats did have to make a "forced" trip. They were "enlisted" by the U.S. Army to eradicate mice. Historian Leo E. Oliva has written about one such occurrence at Fort Atkinson in Kansas during the summer of 1851, when field mice invaded the fort's sod buildings and were destroying provisions. "A requisition for a dozen cats from Fort Leavenworth was filled," he wrote. "So far as is known, this was the first time that cats were carried on the property lists of the army. They did their duty, and the mouse threat to Atkinson was virtually eliminated. Two years later the cats were declared to be 'perfect wrecks' because of another kind of vermin—fleas."[33] 7 - 3 ~ 16 (pm)

Notes

Chapter 1. Buffalo

1. William Becknell, "The Journals of Capt. Thomas [William] Becknell," *Missouri Historical Review* 4 (January 1910): 73.

2. Kate L. Gregg, ed., *The Road to Santa Fe: The Journal and Diaries of George Champlin Sibley* (1952; reprint, Albuquerque: University of New Mexico Press, 1995), 204.

3. Richard Irving Dodge, *The Plains of the Great West and Their Inhabitants* (New York: G. P. Putnam's Sons, 1877), 119.

4. Lydia Spencer Lane, *I Married a Soldier; or Old Days in the Old Army* (1893; reprint, Albuquerque, N.Mex.: Horn & Wallace, 1964), 56.

5. Stanley Vestal, *The Old Santa Fe Trail* (1939; reprint, Lincoln: University of Nebraska Press, 1996), 76.

6. Marc Simmons, "On the Buffalo Trail," Trail Dust (column), *Santa Fe Reporter,* February 3–9, 1993, 24.

7. Ralph P. Bieber, ed., "Diary of Philip Gooch Ferguson, 1847–1848," *Marching with the Army of the West, 1846–1848* (Glendale, Calif.: Arthur H. Clark Co., 1936), 303.

8. Jean Tyree Hamilton, foreword to *Over the Santa Fé Trail in 1857,* by W. B. Napton (Arrow Rock, Mo.: Friends of Arrow Rock, 1991), 4.

9. Napton, *Over the Santa Fé Trail in 1857,* 14.

10. Ibid., 16.

11. Ibid., 17.

12. Ibid.

13. Donald Jackson, ed., *The Journals of Zebulon Montgomery Pike,* vol. 2 (Norman: University of Oklahoma Press, 1966), 343.

14. Ibid.

15. Frank S. Edwards, *A Campaign in New Mexico with Colonel Doniphan* (1847; reprint, Albuquerque: University of New Mexico Press, 1996), 12.

16. Ibid.

17. Dale F. Lott, *American Bison: A Natural History* (Berkeley: University of California Press, 2002), 69.

18. Dodge, *Plains of the Great West,* 120.

19. Ibid.

20. Lott, *American Bison,* 76. For an interesting discourse on bison population, see pp. 69–76 and 167–168.

21. Susan Shelby Magoffin, *Down the Santa Fe Trail and into Mexico: The Diary of Susan Shelby Magoffin, 1846–1847,* ed. Stella M. Drumm (1926; reprint, Lincoln: University of Nebraska Press, 1982), 49.

22. Josiah Gregg, *Commerce of the Prairies,* ed. Max L. Moorhead (1844; reprint, Norman: University of Oklahoma Press, 1954), 374.

23. Albert Pike, *Prose Sketches and Poems, Written in the Western Country* (Boston: Light & Horton, 1834), 63.

24. Lott, *American Bison,* 41–42.

25. Valerius Geist, *Buffalo Nation: History and Legend on the North American Bison* (Stillwater, Minn.: Voyageur Press, 1996), 69.

26. Dodge, *Plains of the Great West,* 110.

27. Ruth Rudner, *A Chorus of Buffalo* (New York: Marlowe & Co., 2003), 114.

28. Ibid., 3–4.

Chapter 2. Pronghorn

1. Gary Turbak, *Pronghorn: Portrait of the American Antelope* (Flagstaff, Ariz.: Northland Press, 1995), 12.

2. Valerius Geist, *Antelope Country; Pronghorns: The Last Americans* (Iola, Wisc.: Krause Publications, 2001), 21.

3. Augustus Storrs and Alphonso Wetmore, *Santa Fe Trail, First Reports: 1825* (Houston, Tex.: Stagecoach Press, 1960), 13.

4. Harry C. Myers, ed., "Alphonso Wetmore's Report of a Journey to Santa Fe in 1828," *Wagon Tracks* 16, no. 4 (August 2002): 9–17.

5. Lewis H. Garrard, *Wah-to-yah and the Taos Trail* (1850; reprint, Norman: University of Oklahoma Press, 1955), 40.

6. Ibid. Garrard used Marcellus St. Vrain, although his correct first name was Marcellin.

7. J. Gregg, *Commerce of the Prairies,* 377–78.

8. Ibid., 378.

9. Napton, *Over the Santa Fé Trail, 1857,* 38, 40.

10. Ibid., 40–41.

11. Joseph Pratt Allyn, *West by Southwest: Letters of Joseph Pratt Allyn, A Traveller along the Santa Fe Trail, 1863,* ed. David K. Strate (Dodge City: Kansas Heritage Center, 1984), 104–105.

12. Magoffin, *Diary of Susan Shelby Magoffin, 1846–1847,* 15.

13. Ibid.

14. Ibid., 50.

15. The given name Marion is used throughout *As Far as the Eye Could Reach,* but "Marian" was used in *Land of Enchantment* by her daughter-in-law Mrs. Hal Russell. The primary reasons for using "Marion" are that the Library of Congress shows her authorized personal name and subject heading as "Russell, Marion Sloan, 1845–1936," and the headstone on her grave at the Stonewall Cemetery in Las Animas County of southeastern Colorado reads "Marion Russell / January 27, 1845 / December 25, 1936 / Mother."

16. Marion Sloan Russell, *Land of Enchantment: Memoirs of Marian [i.e. Marion] Sloan Russell along the Santa Fe Trail / as dictated to Mrs. Hal Russell* (1954; reprint, Albuquerque: University of New Mexico Press, 1981), 22.

17. J. W. Chatham, "Private Journal Commencing February 27, 1849," Center for Southwest Research, University of New Mexico, Albuquerque, 65.

18. William B. Lane, "About Hunting and Snakes," *United Service,* n.s., 11 (April 1894): 317.

19. Marc Simmons, ed., "Ernestine Franke Huning's Diary, 1863," *On the Santa Fe Trail* (Lawrence: University Press of Kansas, 1986), 81–82.

20. Lloyd W. Gundy, ed., "The Journal of Samuel D. Raymond, 1859–1862," *Wagon Tracks* 10, no. 1 (November 1995): 16.

21. Marc Simmons, ed., "Narrative of Hezekiah Brake, 1858," *On the Santa Fe Trail* (Lawrence: University Press of Kansas, 1986), 48.

22. James Francis Riley, "Recollections of James Francis Riley, Part 2," *Wagon Tracks* 9, no. 3 (May 1995): 15–16. The recollections were privately printed in 1959 by John Riley James; a copy was provided to *Wagon Tracks* by Roger F. James.

23. Ibid., 16.

24. J. Gregg, *Commerce of the Prairies,* 378.

25. Jack Schaefer, *An American Bestiary* (Boston: Houghton Mifflin, 1975), [19]–20.

26. Ibid., 20; Turbak, *Pronghorn,* 23–24; Geist, *Antelope Country,* 63–65.

27. Schaefer, *American Bestiary,* 20–21.

28. Ibid., 24; Turbak, *Pronghorn,* 40–43.

29. J. Gregg, *Commerce of the Prairies,* 378.

30. Schaefer, *American Bestiary,* 23; Turbak, *Pronghorn,* 31.

31. Schaefer, *American Bestiary,* 22; Turbak, *Pronghorn,* 19.

32. Twenty-sixth Biennial Pronghorn Workshop 2014, State and Provincial Status Reports for 2013, Western Association of Fish and Wildlife Agencies. This workshop was held at Sul Ross State University, Alpine, Texas, May 12–14. Pronghorn population counts for 2013 were reported at the meetings and provided by James Weaver, district wildlife biologist with the Texas Parks and Wildlife Department, Fort Davis.

33. Twentieth Biennial Pronghorn Workshop 2002. Pronghorn population counts for 2001 were provided by Mace Hack, head of wildlife research for the Nebraska Games and Parks Commission, Lincoln.

34. Geist, *Antelope Country,* 26.

Chapter 3. Prairie Dogs

1. Matthew C. Field, *Matt Field on the Santa Fe Trail,* ed. John E. Sunder (1960; reprint, Norman: University of Oklahoma, 1995), 284.

2. Ibid., 285.

3. Allyn, *West by Southwest,* 61.

4. J. Gregg, *Commerce of the Prairies,* 379.

5. Ibid., 381.

6. William Fairholme, *Journal of an Expedition to the Grand Prairies of the Missouri, 1840,* ed. Jack B. Tykal (Spokane, Wash.: Arthur H. Clark Co., 1996), 117–19.

7. Ibid., 119.

8. Garrard, *Wah-to-yah and the Taos Trail,* 24.

9. Ibid., 23.

10. Bieber, "Diary of Philip Gooch Ferguson, 1847–1848," 305–306.

11. Ibid., 307.

12. Magoffin, *Diary of Susan Shelby Magoffin, 1846–1847,* 37–38. Buffalo meat was indeed available for the Fourth of July dinner, but whether Susan ate anything that day was not mentioned. She was involved in a serious carriage accident that day.

13. Edwards, *A Campaign in New Mexico,* 15–16.

14. W. W. H. [William Watts Hart] Davis, *El Gringo: New Mexico and Her People* (1857; facsimile reprint, Lincoln: University of Nebraska Press, 1982), 41–42.

15. Catherine (Katie) Cary Bowen, *Journal, July 19, 1851,* Bowen Family Papers, United States Army Military History Institute, Carlisle Barracks, Pennsylvania.

16. Donald Jackson, ed., *The Journals of Zebulon Montgomery Pike,* vol. 1 (Norman: University of Oklahoma Press, 1966), 339.

17. Dayton Duncan, *Lewis and Clark: The Journey of the Corps of Discovery* (New York: Alfred A. Knopf, 1997), 40, 84.

18. Jackson, *Journals of Zebulon Montgomery Pike,* 1:338n105, 339n106.

19. W. Eugene Hollon, *The Lost Pathfinder: Zebulon Montgomery Pike* (1949: reprint, Westport, Conn.: Greenwood Press, 1981), 121n1.

20. Jackson, *Journals of Zebulon Montgomery Pike,* 1:338–39.

21. Ibid.

22. J. W. Chatham, "Private Journal Commencing February 27, 1849," Center for Southwest Research, University of New Mexico, Albuquerque, 36.

23. Simmons, "Narrative of Hezekiah Brake, 1858," 46–47.

24. Ibid., 47.

25. James Josiah Webb, *Adventures in the Santa Fé Trade, 1844–1847,* ed. Ralph P. Bieber (1931; reprint, Lincoln: University of Nebraska Press, 1995), 63–65.

26. Gerald Thompson, *Edward F. Beale and the American West* (Albuquerque: University of New Mexico Press, 1983), 113.

27. Dorothy Hinshaw Patent, *Prairie Dogs* (New York: Clarion, 1993), 18.

28. Slobodchikoff, C. N., "Cognition and Communication in Prairie Dogs," in *The Cognitive Animal,* ed. M. Bekoff, C. Allen, and G. M. Burghardt (Cambridge, Mass.: A Bradford Book, MIT Press, 2002), 257–64.

29. Michael E. Long, "The Vanishing Prairie Dog," *National Geographic* 193 (April 1998): 116–30.

30. N. B. Kotliar et al., "A Critical Review of Assumptions about the Prairie Dog as a Keystone Species," *Environmental Management* 24, no. 2 (1999): 177–92.

Chapter 4. Wolves

1. J. Gregg, *Commerce of the Prairies,* 374.

2. E. Lendell Cockrum, *Mammals of the Southwest* (Tucson: University of Arizona Press, 1982), 26.

3. Garrard, *Wah-to-yah and the Taos Trail,* 31–32.

4. Ibid., 27.

5. J. W. Abert, *Abert's New Mexico Report, 1846–47* (1848; reprint, Albuquerque, N.Mex.: Horn & Wallace, 1962), 34.

6. "John James Clemison Diary, Part 1," *Wagon Tracks* 9, no. 3 (May 1995): 9.

7. Marc Simmons, ed., "Trail Letter by Michael Steck, 1852," in *On the Santa Fe Trail* (Lawrence: University Press of Kansas, 1986), 24.

8. William N. Byers, "Letter from Ft. Aubry," *Wagon Tracks* 1, no. 4 (August 1987): 10.

9. Pike, *Prose Sketches and Poems,* 230–31.

10. Barton H. Barbour, ed., *Reluctant Frontiersman: James Ross Larkin on the Santa Fe Trail, 1856–57* (Albuquerque: University of New Mexico Press, 1990), 78.

11. Marc Simmons, ed., "David Kellogg's Diary, 1858," in *On the Santa Fe Trail* (Lawrence: University Press of Kansas, 1986), 55.

12. Garrard, *Wah-to-yah and the Taos Trail,* 18.

13. Robert H. Busch, *The Wolf Almanac* (New York: Lyons & Burford, 1995), 25–26.

14. Edwards, *A Campaign in New Mexico,* 14.

15. J. Gregg, *Commerce of the Prairies,* 375.

16. Marc Simmons, "Wolves Make a 'Howl' of a Subject," New Mexico Scrapbook (column), *Defensor Chieftain* [Socorro, New Mexico], January 10, 1998, 4.

17. J. Gregg, *Commerce of the Prairies,* 375–76.

18. Howard Louis Conard, *"Uncle Dick" Wootton: The Pioneer Frontiersman of the Rocky Mountain Region* (Chicago: W. E. Dibble & Co., 1890), 90.

19. Ibid., 90–91.

20. "Theodore Weichselbaum: Trail Merchant, Part II," *Wagon Tracks* 6, no. 2 (February 1991): 14–15.

21. Leo E. Oliva, *Fort Larned: Guardian of the Santa Fe Trail* (Topeka: Kansas State Historical Society, 1997), 85–88.

22. Simmons, "Narrative by Hezekiah Brake, 1858," 48.

23. George Frederick Augustus Ruxton, *Adventures in Mexico and the Rocky Mountains* (1847; reprint, Glorieta, N.Mex.: Rio Grande Press, 1973), 265.

24. George Frederick Augustus Ruxton, *Life in the Far West,* ed. LeRoy R. Hafen (1849; reprint, Norman: University of Oklahoma Press, 1951), 67.

25. Ruxton, *Adventures in Mexico,* 215.

26. J. Frank Dobie, *The Voice of the Coyote* (Boston: Little, Brown, 1950), xii.

27. Napton, *Over the Santa Fé Trail in 1857,* 52–53.

28. Barbour, *Reluctant Frontiersman,* 76.

29. Edwards, *Campaign in New Mexico,* 14–15.

30. Webb, *Adventures in the Santa Fé Trade, 1844–1847,* 163–64.

31. David E. Brown, ed., *The Wolf in the Southwest: The Making of an Endangered Species* (Silver City, N.Mex.: High-Lonesome Books, 2002), 166.

32. Marc Simmons, "Wolves: Symbol of the Old West," Trail Dust (column), *Santa Fe Reporter,* January 14–20, 1998, 28.

Chapter 5. Coyotes and Roadrunners

1. J. Gregg, *Commerce of the Prairies,* chap. 11, n. 6. The description of the *coyótl,* or coyote, came from Francisco López de Gomara's *Historia Antigua de Méjico,* Vol. 1, chap. 7, n. 17 (original Italian edition, 1552–1553).

2. Ibid.

3. Russell, *Land of Enchantment,* 21.

4. Garrard, *Wah-to-yah and the Taos Trail,* 18.

5. J. Gregg, *Commerce of the Prairies,* 377.

6. Philip St. George Cooke, *Scenes and Adventures in the Army* (Philadelphia: Lindsay & Blakiston, 1857), 269.

7. Ernest Thompson Seton, *Wild Animals at Home* (Garden City, N.Y.: Doubleday, Page & Co., 1917), 12–13.

8. Wyman Meinzer, *Coyote* (Lubbock: Texas Tech University Press, 1995), 8.

9. Dobie, *Voice of the Coyote*, 107.

10. "James Brice's Reminiscences of Ten Years Experience on the Western Plains, Part III," *Wagon Tracks* 7, no. 1 (November 1992): 13.

11. Ruxton, *Life in the Far West*, 66.

12. Ibid., 66–67.

13. Ibid., 181.

14. Vestal, *The Old Santa Fe Trail*, 203.

15. Camilla H. Fox and Christopher M. Papouchis, *Coyotes in Our Midst* (Sacramento, Calif.: Animal Protection Institute, 2005), 5.

16. Vestal, *The Old Santa Fe Trail*, 203.

17. Russell, *Land of Enchantment*, 25.

18. Abert, *Abert's New Mexico Report, 1846–1847*, 149.

19. Ibid., 150.

20. Wyman Meinzer, *The Roadrunner* (Lubbock: Texas Tech University Press, 1993), 10. An excellent source of information about roadrunners, this book includes stunning color photography.

21. Ibid., 70–73. These pages have full-page color photographs showing a pair of roadrunners working together to capture a rattlesnake.

22. J. Gregg, *Commerce of the Prairies*, 138.

23. Marc Simmons, "Roadrunner Lore," New Mexico Scrapbook (column), *Prime Time* [Albuquerque], January 2003, 4.

24. J. Frank Dobie, "The Roadrunner in Fact and Folk-lore," *Texas Ornithological Society Newsletter* 4, no. 4 (May 1, 1956): 1.

25. Ibid., 2.

26. Simmons, "Roadrunner Lore," 4.

27. Meinzer, *Roadrunner*, 6.

Chapter 6. Prairie Chicken

1. Kevin Church, "A Rite of Courtship," *Kansas Wildlife and Parks* (March/April 1989): 2.

2. David Sibley and the National Audubon Society, *The Sibley Guide to Birds* (New York: Alfred A. Knopf, 2000), 147.

3. Ibid.

4. David Sibley et al., *The Sibley Guide to Bird Life and Behavior* (New York: Alfred A Knopf, 2001), 233.

5. Kevin J. Zimmer, *The Western Bird Watcher* (Englewood Cliffs, N.J.: Prentice Hall, 1985), 79.

6. Ted T. Cable *et al.*, *Birds of Cimarron National Grassland,* Gen. Tech. Report RM-GTR-281 (Fort Collins, Colo.: U.S. Department of Agriculture, Forest Service, 1996), 45.

7. Ibid.

8. Roger Tory Peterson, *A Field Guide to Western Birds* (Boston: Houghton Mifflin, 1990), 160; Sibley and National Audubon Society, *Sibley Guide to Birds,* 149.

9. Sibley and National Audubon Society, *Sibley Guide to Birds,* 233.

10. Rob Manes, "Phantoms of the Prairie," *Kansas Wildlife* (November/December 1983), 19.

11. William Least Heat-Moon, *PrairyErth* (Boston: Houghton Mifflin, 1991), 82.

12. Gerald J. Horak, "The Prairie Bird," *Kansas Wildlife* (November/December 1987): 12.

13. Peterson, *Field Guide to Western Birds,* 160.

14. Hal H. Harrison, *Western Birds' Nests* (Boston: Houghton Mifflin, 1979), 50–51.

15. Russell, *Land of Enchantment*, 22.

16. J. Gregg, *Commerce of the Prairies*, 383.

17. Ibid.

18. Simmons, "David Kellogg's Diary, 1858," 55.

19. Lloyd W. Gundy, "The Journal of Samuel D. Raymond, 1859–1862," *Wagon Tracks* 10, no. 1 (November 1995): 13.

20. Garrard, *Wah-to-yah and the Taos Trail*, 13.

21. Simmons, "Trail Letter by Michael Steck, 1852," 21.

22. Ibid., 22.

23. J. W. Chatham, "Private Journal Commencing February 27, 1849," Center for Southwest Research, University of New Mexico, Albuquerque, 35.

24. Marc Simmons and Hal Jackson, *Following the Santa Fe Trail,* 3rd ed. (Santa Fe, N.Mex.: Ancient City Press, 2001), 75. There were two routes from Fort Leavenworth connecting to the Santa Fe Trail; one route crossed the Kansas River near its confluence with the Wakarusa River in present-day Douglas County.

25. Leo E. Oliva, ed., "'A Faithful Account of Everything': Letters from Katie (Catherine Cary) Bowen on the Santa Fe Trail, 1851," *Kansas History* 19, no. 4 (Winter 1996–1997): 270.

26. Ibid., 281.

27. James Francis Riley, "Recollections of James Francis Riley, 1838–1918, Part 1," *Wagon Tracks* 9, no. 2 (February 1995): 16.

28. Samuel P. Arnold, *Eating Up the Santa Fe Trail* (Niwot: University Press of Colorado, 1990). This is the only book published about the foods and recipes from the time of the old Trail. Sam Arnold was interviewed in 2002; he passed away in 2006.

29. Horak, "The Prairie Bird," 10.

30. Bill Van Pelt, "Aerial Survey Shows Lesser Prairie Chicken Population Increased 20 Percent in 2014," Western Association of Fish and Wildlife Agencies, July 1, 2014, http://www.wafwa.org.

Chapter 7. Rattlesnakes

1. Laurence M. Klauber, *Rattlesnakes: Their Habits, Life Histories and Influence on Mankind* (Berkeley: University of California Press, 1982), 5–8.

2. Ibid., 105–10.

3. J. Frank Dobie, *Rattlesnakes* (Boston: Little, Brown, 1965), 160.

4. Marc Simmons, "Some Rattlesnake Lore," Trail Dust (column), *Santa Fe Reporter,* November 24–30, 1993, 17.

5. Ibid.

6. Edwards, *A Campaign in New Mexico*, 16.

7. Simmons, "Some Rattlesnake Lore," 17.

8. David A. White, ed., *News of the Plains and Rockies, 1803–1865*, Vol. 2, *Santa Fe Adventurers, 1818–1843* (Spokane, Wash.: Arthur H. Clark, 1998), 62–63.

9. K. Gregg, *Journal and Diaries of George Champlin Sibley*, 65.

10. Ibid.

11. J. Gregg, *Commerce of the Prairies*, 46.

12. Ibid.

13. Ibid., 382.

14. Magoffin, *Diary of Susan Shelby Magoffin, 1846–1847*, 50.

15. John W. Moore, "Trail Trip, 1867," *Wagon Tracks* 4, no. 2 (February 1990): 18.

16. Donald R. Hale, "The Old Plainsmen's Association," ed. Mark L. Gardner, *Wagon Tracks* 14, no. 3 (May 2000): 18.

17. Simmons, "Some Rattlesnake Lore," 17.

18. James Brice, "James Brice's Trail Reminiscences, Part II," *Wagon Tracks* 6, no. 4 (August 1992): 12.

19. Simmons, "David Kellogg's Diary, 1858," 56.

20. W. B. Lane, "About Hunting and Snakes," 317–22.

21. Ibid.

22. Ibid.

23. Bob Myers, "How Dangerous Are Rattlesnakes?" (Albuquerque, N.Mex.: American International Rattlesnake Museum, 1991), 1.

24. Bieber, "Diary of Philip Gooch Ferguson, 1847–1848," 311.

25. L. S. Lane, *I Married a Soldier*, 25.

26. Ibid., 42.

27. Field, *Matt Field on the Santa Fe Trail*, 284.

28. Hale, "The Old Plainsmen's Association," 18.

29. Russell, *Land of Enchantment*, 67.

30. Dobie, *Rattlesnakes*, 152.

31. Ibid., 151.

Chapter 8. Grizzlies and Black Bears

1. Marc Simmons, "Get the Grizzly!" Trail Dust (column), *Santa Fe Reporter*, January 8, 1986, 12.

2. Robert Busch, *The Grizzly Almanac* (New York: Lyons Press, 2000), 27.

3. J. Gregg, *Commerce of the Prairies*, 377.

4. Ibid., 136.

5. James O. Pattie, *The Personal Narrative of James O. [Ohio] Pattie of Kentucky*, ed. Timothy Flint (1831; reprint, Chicago: Lakeside Press, 1930), 180.

6. Ibid., 180–181.

7. "John James Cleminson Diary: Part 1," *Wagon Tracks* 9, no. 3 (May 1995): 9.

8. Allyn, *West by Southwest*, 107.

9. Duncan, *Lewis and Clark*, [38].

10. Meriwether Lewis and William Clark, *The Journals of Lewis and Clark*, ed. Anthony Brandt (Washington, D.C.: National Geographic Society, 2002), 120–21.

11. Ibid., 138.

12. Ibid., 146.

13. Ibid., 156.

14. Jackson, *Expeditions of Zebulon Montgomery Pike*, 2: 283–84.

15. Ibid., 294.

16. Ibid.

17. Busch, *Grizzly Almanac*, 25.

18. Conard, *"Uncle Dick" Wootton*, 157–58.

19. Colonel Henry Inman, *The Old Santa Fé Trail: The Story of a Great Highway* (1897; reprint, Williamstown, Mass.: Corner House Publishers, 1977), 30–37. Refer to these pages for Inman's full-length version of Williams's narrative, which is condensed for this chapter. For the original presentation of Williams's adventures, see David H. Coyner, *The Lost Trappers*, ed. David J. Weber (1847; reprint, Albuquerque: University of New Mexico Press, 1970), 66–86.

20. George Frederick Augustus Ruxton, *Adventures in Mexico and the Rocky Mountains* (1847; reprint, Glorieta, N.M.: Rio Grande Press, 1973), 270.

21. Inman, *The Old Santa Fé Trail*, 206–207.

22. Pattie, *Personal Narrative*, 35–39, 51.

23. Ruxton, *Adventures in Mexico and the Rocky Mountains*, 270–273. Ruxton tells a shortened version of the saga of Hugh Glass, referring to him as John Glass. Several books have been published about Glass and his legendary ordeal, including Fredrick Manfred's *Lord Grizzly* (1954) and Bruce Bradley's *Hugh Glass* (1999).

24. Simmons, "Get the Grizzly!" 12.

25. David E. Brown, *The Grizzly in the Southwest: Documentary of an Extinction* (Norman: University of Oklahoma Press, 1996), 4.

26. Busch, Robert H., "Epilogue," *Valley of the Grizzlies* (New York: St. Martin's Press, 1998), 111.

Chapter 9. Mustangs

1. J. Gregg, *Commerce of the Prairies*, 365.

2. Ibid., 366.

3. Pike, *Prose Sketches and Poems*, 63.

4. J. Frank Dobie, *The Mustangs* (New York: Bramhall House, 1952), xii.

5. Ibid., xv–xvi.

6. Alvin and Virginia Silverstein, *The Mustang* (Brookfield, Conn.: Millbrook Press, 1997), 14–15.

7. Hope Ryden, "The Colors of Horses," *Wild Horses I Have Known* (New York: Clarion, 1999), 82-[88]. For descriptions of the colors of horses (a fascinating and complex subject in itself), see the website of the American Paint Horse Association and the books of Ben K. Green.

8. J. Gregg, *Commerce of the Prairies*, 366.

9. Vestal, *Old Santa Fe Trail*, 23–24.

10. Allyn, *West by Southwest*, 143.

11. Vestal, *Old Santa Fe Trail*, [21]–22.

12. Ibid., 24.

13. Garrard, *Wah-to-yah and the Taos Trail*, 10.

14. Ibid., 11.

15. Ibid., 16.

16. Ibid., 15.

17. Dobie, *Mustangs*, 107.

18. Vestal, *Old Santa Fe Trail*, [21]–22.

19. Dan Flores, "The Horse Nations Endure," in *Unbroken Spirit: The Wild Horse in the American Landscape,* ed. Frances B. Clymer and Charles R. Preston (Cody, Wyo.: Buffalo Bill Historical Center, 1999), 43.

20. Schaefer, *American Bestiary,* 21.

21. Dobie, *Mustangs,* 21; John Stephen Hockensmith, *Spanish Mustangs in the Great American West: Return of the Horse* (Norman: University of Oklahoma Press, 2009), 24.

22. Dobie, *Mustangs,* 25.

23. Ibid., 42.

24. Marc Simmons, "Mustang Days," Trail Dust (column), *Santa Fe Reporter,* November 8–14, 1995, 29.

25. Flores, "Horse Nations Endure," 43.

26. J. Gregg, *Commerce of the Prairies,* 367.

27. Jackson, *Journals of Zebulon Montgomery Pike,* 2:340–41.

28. Ibid., 342.

29. William Y. Chalfant, *Dangerous Passage: The Santa Fe Trail and the Mexican War* (Norman: University of Oklahoma Press, 1994), 56.

30. Jackson, *Journals of Zebulon Montgomery Pike,* 2:342.

31. Dobie, *Mustangs,* 119.

32. Washington Irving, *A Tour on the Prairies* (London: Henry G. Bohn, 1850), 69.

33. K. Gregg, *Journal and Diaries of George Champlin Sibley,* 75.

34. Pike, *Prose Sketches and Poems,* 9.

35. Ibid., 63–64.

36. Barton H. Barbour, ed., *Reluctant Frontiersman, James Ross Larkin on the Santa Fe Trail, 1856–1857* (Albuquerque: University of New Mexico Press, 1990), 80.

37. Russell, *Land of Enchantment,* 81.

38. Sandra L. Myres, ed., *Cavalry Wife: The Diary of Eveline M. Alexander, 1866–1867* (College Station: Texas A&M University Press, 1977), 79.

39. Ibid.

40. George E. Hyde, *Life of George Bent, Written from His Letters,* ed. Savoie Lottinville (Norman: University of Oklahoma Press, 1968), 37.

41. Ibid., 35.

42. Garrard, *Wah-to-yah and the Taos Trail,* 53–54.

43. Dobie, *Mustangs,* 199.

44. J. Gregg, *Commerce of the Prairies,* 324n12.

Chapter 10. Oxen

1. Riley, "Recollections, Part 1," 15.

2. "Oxen Questions Most Asked," *The Prairie Ox Drovers,* http://www.prairie oxdrovers.com.

3. Marc Simmons, "Little-known Ox Lore Gives Animals Their Due," Trail Dust (column), *Santa Fe New Mexican,* August 27, 2005, C3.

4. Inman, *Old Santa Fé Trail,* 87. See also "The 1829 Escorts," by Leo E. Oliva, in *Confrontation on the Santa Fe Trail: Selected Papers from Santa Fe Trail Association Symposia, 1993 and 1995,* ed. Leo E. Oliva (Woodston, Kan.: Santa Fe Trail Association Publications, 1996), 17–24; and Otis E. Young, *The First Military Escort on the Santa Fe Trail, 1829,* (Glendale, Calif.: Arthur H. Clark Co., 1952).

5. J. Gregg, *Commerce of the Prairies,* 25.

6. Ibid., 24.

7. David K. Clapsaddle, "Old Dan and His Traveling Companions: Oxen on the Santa Fe Trail," *Wagon Tracks* 22, no. 2 (February 2008): 10.

8. Ibid.

9. Henry Walker, *The Wagonmasters* (Norman: University of Oklahoma Press, 1966), 106–108.

10. J. Gregg, *Commerce of the Prairies,* 73.

11. Ibid., 72–73.

12. Riley, "Recollections, Part 1," 15, 21.

13. Franz Huning, *Trader on the Santa Fe Trail: Memoirs of Franz Huning. With Notes by His Granddaughter, Lina Fergusson Browne* (Albuquerque, N.Mex.: University of Albuquerque, Calvin Horn Publisher, 1973), 10. The University of Albuquerque closed in the late 1980s. It was not part of the University of New Mexico.

14. Ibid., 13.

15. Ibid., 17.

16. David K. Clapsaddle, preface to *Rules and Regulations by Which to Conduct Wagon Trains (Drawn by Oxen on the Plains),* by Tom C. Cranmer (1866; reissue, Larned, Kan.: Wet/Dry Routes Chapter, Santa Fe Trail Association, 2007), [p. 1].

17. Cranmer, *Rules and Regulations,* [p. 4].

18. Ibid., 8–9.

19. Ibid., 22.

20. Ibid., 10–11.

21. Magoffin, *Diary of Susan Shelby Magoffin, 1846–1847,* 9.

22. Marc Simmons, "Bullwhacking," Trail Dust (column), *Santa Fe Reporter,* March 11–17, 1998, 27.

23. George E. Vanderwalker, "Reminiscences, 1864," in *On the Santa Fe Trail,* ed. Marc Simmons (Lawrence: University Press of Kansas, 1986), 94.

24. Ibid, 88–89.

25. Simmons, "Bullwhacking," 27.

26. Russell, *Land of Enchantment,* 17–18.

27. Marc Simmons, "Oxen versus Mules," in *The Old Trail to Santa Fe: Collected Essays* (Albuquerque: University of New Mexico Press, 1996), 143.

28. Walker, *The Wagonmasters,* 109–10.

29. Donald R. Hale, ed. Mark L. Gardner, "Old Plainsmen's Association," *Wagon Tracks* 14, no. 3 (May 2000): 18.

30. Matt. 11:28–30, King James Version.

Chapter 11. Mules

1. Max L. Moorhead, "Spanish Transportation in the Southwest, 1540–1846," *New Mexico Historical Review* 32, no. 2 (April 1957), 108–109.

2. Marc Simmons, "The Mule: An Unsung Hero of the Desert Southwest," Trail Dust (column), *Santa Fe New Mexican,* November 15, 2003, B1.

3. Floyd F. Ewing, Jr., "The Mule as a Factor in the Development of the Southwest," *Arizona and the West* 5, no. 4 (Winter 1963): 315.

4. "What Is a Mule?" The American Donkey and Mule Society, Lewisville, Texas, http://www.lovelongears.com.

5. Marc Simmons and Hal Jackson, *Following the Santa Fe Trail,* 3rd ed., revised and expanded (Santa Fe, N.Mex.: Ancient City Press, 2001), 1.

6. Jack D. Rittenhouse, *Trail of Commerce and Conquest* (1971; reprint, Woodston, Kan.: Santa Fe Trail Association, 2000), 9.

7. Robert L. Duffus, *The Santa Fe Trail* (1930; reprint, Albuquerque: University of New Mexico Press, 1972), 81.

8. Rittenhouse, *Trail of Commerce and Conquest,* 11.

9. Duffus, *Santa Fe Trail,* 108.

10. Ibid., 133.

11. J. Gregg, *Commerce of the Prairies* (1844; reprint, Norman: University of Oklahoma Press, 1954), 127.

12. Marc Simmons, "Mule Drivers Were the First Convoys of the Southwest Roads," History (column), *New Mexico Independent* (Albuquerque), March 31, 1978, 8.

13. Marc Simmons, "In Praise of the Frontier Mule," Trail Dust (column), *Santa Fe Reporter,* September 1–7, 1993, 21.

14. Ibid.

15. Magoffin, *Diary of Susan Shelby Magoffin, 1846–1847,* 2–3.

16. Ibid., 3.

17. Anna Belle Cartwright, ed., "William James Hinchey, An Irish Artist on the Santa Fe Trail, Part I." *Wagon Tracks* 10, no. 3 (May 1996): 21. See the bibliography of this book for the citation with information about all three parts of this essay.

18. Ibid.

19. Vestal, *Old Santa Fe Trail,* 46.

20. Webb, *Adventures in the Santa Fé Trade, 1844–1847,* 62.

21. Ibid., 60. Webb refers to Marcellin St. Vrain, his correct name. Lewis Garrard used Marcellus.

22. Statistics on mules provided by the U.S. Department of Agriculture, National Agricultural Statistics Service, Livestock Section, "Statistics of Agriculture, Miscellaneous Live Stock, Mules," table 11, June 1, 1890, http://www.nass.usda.gov.

23. Sonie Liebler, "Steamboat Arabia," *Wagon Tracks* 9, no. 4 (August 1995): 1, 7–9. The Treasures of the Steamboat *Arabia* Museum is a "must-see" when visiting Kansas City, Missouri.

Chapter 12. Burros (Donkeys) and Horses

1. Susan L. Woodward, "The Living Relatives of the Horse," chap. 10 in *Horses through Time*, ed. Sandra L. Olsen (Boulder, Colo.: Roberts Rinehart Publishers for Carnegie Museum of Natural History, 1995), 202.

2. K. Gregg, *Journal and Diaries of George Champlin Sibley, 1825–1827*, 215.

3. J. Gregg, *Commerce of the Prairies*, 133.

4. Ibid., 135.

5. Allyn, *West by Southwest*, 99.

6. Frank Brookshier, *The Burro* (Norman: University of Oklahoma Press, 1974), 265–66.

7. Davis, *El Gringo*, 173–74.

8. Ibid., 174.

9. Richard Rudisill and Marcus Zafarano, *Burros* (Santa Fe: Museum of New Mexico, 1979), n.p.

10. Marc Simmons, "Mexican Burro Representative of Southwest Trails," Trail Dust (column), *Santa Fe New Mexican*, March 10, 2001, B1, B3.

11. Ibid.

12. Interview with Marc Simmons, San Marcos Cafe, Cerrillos, New Mexico, May 26, 2011.

13. Ibid.

14. Brookshier, *Burro*, 15.

15. Woodard, *Horses through Time*, 204, 208.

16. Les Vilda, "A Modern Encounter with the Trail," *Wagon Tracks* 3, no. 3 (May 1989): 6.

17. Ibid.

18. Vestal, *Old Santa Fe Trail*, 21–22.

19. K. Gregg, *Road to Santa Fe*, 34–76, 105–11.

20. Napton, *Over the Santa Fé Trail, 1857*, 10.

21. Alexander, *Cavalry Wife*, 34–35.

22. Ibid., 97.

23. Allyn, *West by Southwest*, 143.

24. Russell, *Land of Enchantment*, 23.

25. James Francis Riley, "Recollections of James Francis Riley, Part III," *Wagon Tracks* 9, no. 4 (August 1995): 21.

26. "Historic Domestic Animal Statistics," U.S. Department of Agriculture, National Agricultural Statistics Service, Livestock Section, http://www.nass.usda.gov. Also, for information about donkeys, see "All About Donkeys!" http://www.lovelongears.com.

Chapter 13. Dogs

1. Marc Simmons, "New Mexico Historians Have Forgotten the Role of the Dog," *New Mexico Independent* (Albuquerque), December 1, 1978, 9.

2. L. S. Lane, *I Married a Soldier*, 25–26.

3. Cranmer, *Rules and Regulations*, 26.

4. Washington Irving, *A Tour on the Prairies* (London: John Murray, Albemarle Street, 1835), 40–41.

5. Field, *Matt Field on the Santa Fe Trail*, 55.

6. Ibid., 309.

7. Ibid., 309–10.

8. Ibid., 310–11.

9. Magoffin, *Diary of Susan Shelby Magoffin, 1846–1847*, 4.

10. Ibid., 13–14.

11. Ibid., 18.

12. Ibid., 30.

13. Ibid., 49.

14. Ibid., 78.

15. Catherine (Katie) Cary Bowen Letters, ed. Leo E. Oliva, *Wagon Tracks* 16, no. 4 (August 2002): 23. Katie Bowen's letters appear in *Wagon Tracks* beginning in vol. 16, no. 2 (February 2002).

16. Bowen, "A Faithful Account of Everything," 271–72.

17. Ibid., 274.

18. Mark Derr, *A Dog's History of America* (New York: North Point Press, 2004), 175–76.

19. Ibid., 176.

20. Ibid.

21. David Lavender, *Bent's Fort* (1954; reprint, Lincoln: University of Nebraska Press, 1972), 131–33.

22. Ibid., 257.

23. Ibid., 257–58.

24. Garrard, *Wah-to-yah and the Taos Trail*, 131.

25. Marc Simmons, "Sheep Dogs in the Southwest," Trail Dust (column), *Santa Fe Reporter*, July 11–17, 1990, 15.

26. Garrard, *Wah-to-yah and the Taos Trail*, 61.

27. Ibid., 66.

28. "Theodore Weichselbaum: Trail Merchant, Part II," *Wagon Tracks* 6, no. 2 (February 1992): 13.

29. Miguel Antonio Otero, Jr., *My Life on the Frontier, 1864–1882* (1935; reprint, Albuquerque: University of New Mexico Press, 1987), 80.

30. Ibid., 80–81.

31. Ibid., 81.

32. Ibid.

33. Leo E. Oliva, "Fort Atkinson on the Santa Fe Trail, 1850–1854," *Kansas Historical Quarterly* (Summer 1974; reprint, Wet/Dry Routes Chapter, Santa Fe Trail Association, Larned, Kan., 2007), 11–12.

Bibliography

This bibliography comprises about three hundred citations to sources that I read or examined and used in writing this book. The first section details reference sources about the famous Trail. The second, Primary Sources, lists works of travelers who wrote about their experiences along its routes. Under the third section heading, The Santa Fe Trail, are four subsections exploring sources related to its history, trade, and many other aspects. Included are guidebooks, photography and art, and websites. The fourth and last section, The Animals, is devoted to the species included in this book.

Bibliographical Works

1971   Jack D. Rittenhouse. *The Santa Fe Trail: A Historical Bibliography*. Albuquerque: University of New Mexico Press. 271 pages. Published in 1971 to commemorate the 150th anniversary of the opening of the Santa Fe Trail by William Becknell, this reference book is a briefly annotated listing of 718 titles related to this historic "Highway of Commerce." It is an important resource for researchers and others interested in the literature of the Trail.

1982   Henry R. Wagner and Charles L. Camp. *The Plains and the Rockies: A Critical Bibliography of Exploration, Adventure, and Travel in the American West, 1800–1865*. Fourth ed., revised and enlarged. Edited by Robert H. Becker. San Francisco: John Howell Books.

1986   Marc Simmons. "Women on the Santa Fe Trail: Diaries, Journals, Memoirs; An Annotated Bibliography." *New Mexico Historical Review* 61, no. 3 (July 1986): 233–43.

2005   Phyllis S. Morgan. *Marc Simmons of New Mexico: Maverick Historian*. Foreword by Mark L. Gardner. Albuquerque: University of New Mexico Press. Part I of this bio-bibliography is a biography about Marc Simmons, historian, author, and recognized authority on the Santa Fe National Historic Trail.

Part II is a comprehensive bibliography of Simmons's works from 1965 to 2005, comprising more than three thousand citations, many of which are annotated. His principal subjects include the history of New Mexico and the American Southwest, the history of the Santa Fe Trail and its myriad subjects, and Spanish Colonial history of the Southwest. His history columns in newspapers have included writings about a variety of wild and domestic animals.

2006 Harry C. Myers. "Since Rittenhouse: Santa Fe Trail Bibliography." *Wagon Tracks* 20, no. 4 (August 2006): 20–29. This issue commemorated the twentieth anniversary of *Wagon Tracks,* the quarterly of the Santa Fe Trail Association.

Primary Sources

Abert, J. W. *Abert's New Mexico Report, 1846–1847.* 1848. Facsimile reprint with a foreword by William A. Keleher. Albuquerque, N.Mex.: Horn & Wallace, 1962.
———. *Western America in 1846–1847: The Original Travel Diary of Lieutenant J. W. Abert.* Edited by John Galvin. San Francisco: John Howell Books, 1966.
Alexander, Eveline M. *Cavalry Wife: The Diary of Eveline M. Alexander, 1866–1867.* Edited by Sandra L. Myres. College Station: Texas A&M University Press, 1977.
Allyn, Joseph Pratt. *West by Southwest: Letters of Joseph Pratt Allyn, A Traveller along the Santa Fe Trail, 1863.* Edited by David K. Strate. Dodge City: Kansas Heritage Center, 1984.
Becknell, William. "The Journals of Capt. Thomas [William] Becknell from Boone's Lick to Santa Fe and from Santa Cruz to Green River." *Missouri Historical Review* 4 (January 1910): 65–84.
Bent, Charles. "The Charles Bent Papers." Parts 1–8. Edited by Frank D. Reeve. *New Mexico Historical Review* 29 (July, October 1954): 234–39, 311–17; 30 (April, July, October 1955): 154–67, 252–54, 340–52; 31 (January, April, July 1956): 75–77, 157–64, 251–53.
Boggs, William M. "The W. M. Boggs Manuscript about Bent's Fort, Kit Carson, the Far West, and Life among the Indians." Edited by LeRoy R. Hafen. *Colorado Magazine* 7 (March 1930): 45–69.
Bowen, Catherine (Katie) Cary. "'A Faithful Account of Everything': Letters from Katie Bowen on the Santa Fe Trail, 1851." Edited by Leo E. Oliva. *Kansas History* 19, no. 4 (Winter 1996–1997): 262–81.
———. "Letters." Edited by Leo E. Oliva. *Wagon Tracks* 16, nos. 2–4 (February–August 2002).
Brake, Hezekiah. "Narrative by Hezekiah Brake, 1858." In *On the Santa Fe Trail,* edited by Marc Simmons, 37–51. Lawrence: University Press of Kansas, 1986.

Brice, James. "James Brice's Trail Reminiscences of Ten Years Experience on the Western Plains." Parts 1–3. *Wagon Tracks* 6, no 3 (May 1992): 1, 19–21; 6, no. 4 (August 1992): 10–13; 7, no. 1 (November 1992): 12–15.

Byers, William N. "Letter from Ft. Aubry." *Wagon Tracks* 1, no. 4 (August 1987): 10.

Carson, Kit. *Kit Carson's Autobiography.* 1935. Facsimile reprint edited by Milo Milton Quaife. Lincoln: University of Nebraska Press, 1966.

Chatham, J. W. "Private Journal of J. W. Chatham Commencing February 27, 1849." Center for Southwest Research, Zimmerman Library, University of New Mexico, Albuquerque.

Cleminson, John James. "John James Cleminson Diary: Part 1." *Wagon Tracks* 9, no. 3 (May 1995): 9.

Conard, Howard Louis. *"Uncle Dick" Wootton: The Pioneer Frontiersman of the Rocky Mountain Region.* Introduction by Joseph Kirkland. Chicago: W. E. Dibble & Co. 1890. Facsimile reprint, Alexandria, Va.: Time-Life, 1980.

Connelley, William Elsey. *War with Mexico, 1846–1847: Doniphan's Expedition and the Conquest of New Mexico and California.* 1907. Facsimile reprint, Bowie, Md.: Heritage Books, 2000.

Cooke, Philip St. George. *Scenes and Adventures in the Army: Or, Romance of Military Life.* Philadelphia: Lindsay & Blakiston, 1859.

Davis, W. W. H. (William Watts Hart). *El Gringo: New Mexico and Her People.* 1857. Facsimile reprint, Lincoln: University of Nebraska Press, 1982.

Edwards, Frank S. *A Campaign in New Mexico with Colonel Doniphan.* Philadelphia: Carey & Hart, 1847. Reprinted with a foreword by Mark L. Gardner. Albuquerque: University of New Mexico Press, 1996.

Elliott, Richard Smith. *The Mexican War Correspondence of Richard Smith Elliott.* Edited and annotated by Mark L. Gardner and Marc Simmons. Norman: University of Oklahoma Press, 1997.

Fairholme, William. *Journal of an Expedition to the Grand Prairies of the Missouri, 1840.* 1841. Reprint edited by Jack B. Tykal. Spokane, Wash.: Arthur H. Clark Co., 1996.

Ferguson, Philip Gooch. "Diary of Philip Gooch Ferguson, 1847–1848." *Marching with the Army of the West, 1846–1848.* Edited by Ralph P. Bieber. Glendale, Calif.: Arthur H. Clark Co., 1936.

Field, Matthew C. "The Diary of Matt Field." Edited by William G. B. Carson. *Bulletin of the Missouri Historical Society* 5 (April 1949): 157–84.

———. *Matt Field on the Santa Fe Trail.* Collected by Clyde and Mae Reed Porter. Edited and with an introduction and notes by John E. Sunder. 1960. Reprinted with a foreword by Mark L. Gardner. Norman: University of Oklahoma Press, 1995.

Fowler, Jacob. *The Journal of Jacob Fowler.* Edited by Elliott Coues. New York: Harper, 1898. Reprinted with a preface and notes by Raymond W. Settle, Mary Lund Settle, and Harry R. Stevens. Lincoln: University of Nebraska Press, 1970.

Garrard, Lewis H. *Wah-to-yah and the Taos Trail.* 1850. Reprinted with an introduction by A. B. Guthrie, Jr. Norman: University of Oklahoma Press, 1955.

Gibson, George Rutledge. *Over the Chihuahua and Santa Fe Trails, 1847–1848: George Rutledge Gibson's Journal.* Edited and annotated by Robert W. Frazer. Albuquerque: University of New Mexico Press, 1981.

Glasgow, Edward James, and William Henry Glasgow. *Brothers on the Santa Fe and Chihuahua Trails: Edward James Glasgow and William Henry Glasgow, 1846–1848.* Edited and annotated by Mark L. Gardner, with a foreword by Marc Simmons. Niwot: University Press of Colorado, 1993.

Gregg, Josiah. *Commerce of the Prairies.* 1844. Reprint edited by Max L. Moorhead, with a foreword by Marc Simmons. Norman: University of Oklahoma Press, 1954.

Gregg, Kate L., ed. *The Road to Santa Fe: The Journal and Diaries of George Champlin Sibley and Others Pertaining to the Surveying and Marking of a Road from the Missouri Frontier to the Settlements of New Mexico, 1825–1827.* Albuquerque: University of New Mexico Press, 1995.

Hardesty, George W. "Diary of George W. Hardesty." Edited by Richard H. Louden. *Wagon Tracks* 9, no. 2 (February 1995): 11.

Hinchey, William James. "William James Hinchey, An Irish Artist on the Santa Fe Trail." Edited by Anna Belle Cartwright. Parts 1–3. *Wagon Tracks* 10, no. 3 (May 1996): 11–15, 19–23; 10, no. 4 (August 1996): 12–22; 11, no. 1 (November 1996): 10–18.

Hughes, John Taylor. *Doniphan's Expedition, Containing an Account of the Conquest of New Mexico.* Cincinnati, Ohio: U. P. James, 1847.

Hulbert, Archer Butler, ed. *Southwest on the Turquoise Trail.* Denver, Colo.: Denver Public Library, 1933. A collection of short diaries from the earliest years (1820s) of the Santa Fe Trail.

Huning, Ernestine Franke. "Ernestine Franke Huning's Diary, 1863." In *On the Santa Fe Trail,* edited by Marc Simmons, 73–83. Lawrence: University Press of Kansas, 1986.

Huning, Franz. *Trader on the Santa Fe Trail: Memoirs of Franz Huning. With Notes by His Granddaughter, Lina Fergusson Browne.* Albuquerque, N.Mex.: University of Albuquerque Press, Calvin Horn Publisher, 1973.

Hyde, George E. *Life of George Bent, Written from His Letters.* Edited by Savoie Lottinville. Norman: University of Oklahoma Press, 1968.

Kellogg, David. "David Kellogg's Diary, 1858." In *On the Santa Fe Trail,* edited by Marc Simmons, 52–63. Lawrence: University Press of Kansas, 1986.

Kingsbury, John M. *Trading in Santa Fe: John M. Kingsbury's Correspondence with James Josiah Webb, 1853–1861.* Edited by Jane Lenz Elder and David J. Weber. Dallas, Tex.: Southern Methodist University Press, 1996.

Lane, Lydia Spencer. *I Married a Soldier; or, Old Days in the Old Army.* 1893. Reprint, Albuquerque: Horn & Wallace, 1964.

Lane, William B. "About Hunting and Snakes." *United Service,* n.s., 11 (April 1984): 317–22.

Larkin, James Ross. *Reluctant Frontiersman: James Ross Larkin on the Santa Fe Trail.* Edited and annotated by Barton H. Barbour. Albuquerque: University of New Mexico Press, 1990.

Magoffin, Susan Shelby. *Down the Santa Fe Trail and into Mexico: The Diary of Susan Shelby Magoffin, 1846–1847.* Edited by Stella M. Drumm. New Haven, Conn.: Yale University Press, 1926. Reprinted with a foreword by Howard R. Lamar. Lincoln: University of Nebraska Press, 1982.

Marmaduke, Meredith Miles. "Santa Fe Trail: M. M. Marmaduke Journal (1824)." *Missouri Historical Review* 6, no. 1 (October 1911): 1–10.

Meyer, Marian. *Mary Donoho: New First Lady of the Santa Fe Trail.* With a foreword by Marc Simmons. Santa Fe, N.Mex.: Ancient City Press, 1991.

Napton, W. B. (William B., Jr.) *Over the Santa Fé Trail in 1857.* 1905. Reprint of the first edition with a foreword by Jean Tyree Hamilton. Arrow Rock, Mo.: Friends of Arrow Rock, 1991.

New, David F. *Tales of the Trail—Mexican War, 1846–1849; The Personal Correspondence of Soldiers and Men Who Made the Long Journey to Santa Fe in 1846 to 1849.* Mount Vernon, Wash.: David F. New, 2013. Forty letters of soldiers serving volunteer army units.

Otero, Miguel Antonio. *My Life on the Frontier, 1864–1882.* Illustrated by Will Shuster. 2 vols. New York: Press of the Pioneers, 1935–1939. Facsimile reprint, Albuquerque: Sunstone Press, 2007.

Pattie, James Ohio. *The Personal Narrative of James O. Pattie of Kentucky.* Edited by Timothy Flint. 1831. Reprinted with historical introduction and notes by Milo Milton Quaife. Chicago: Lakeside Press, 1930.

Pike, Albert. *Prose Sketches and Poems Written in the Western Country.* First ed. Boston: Light & Horton, 1834.

Pike, Zebulon Montgomery. *The Journals of Zebulon Montgomery Pike, with Letters and Related Documents.* Edited and annotated by Donald Jackson. 2 vols. Norman: University of Oklahoma Press, 1966.

Raymond, Samuel D. "The Journal of Samuel D. Raymond, 1859–1862." Transcribed, annotated, and introduced by Lloyd W. Gundy. *Wagon Tracks* 10, no. 1 (November 1995): 1, 12–20.

Riley, James Francis. "Recollections of James Francis Riley, 1838–1918." Parts 1 and 2. *Wagon Tracks* 9, no. 2 (February 1995): 13–22; 9, no. 3 (May 1995): 11–21. Privately printed by John Riley James, and copy provided for publication in *Wagon Tracks* by Roger F. James.

Russell, Marion Sloan. *Land of Enchantment: Memoirs of Marian [i.e. Marion] Sloan Russell along the Santa Fe Trail / as dictated to Mrs. Hal Russell.* Edited by Garnet M. Bryer. Evanston, Ill.: Branding Iron Press, 1954. An illustrated facsimile of the first edition, with an afterword by Marc Simmons, was printed by the University of New Mexico Press in 1981.

Ruxton, George Frederick Augustus. *Adventures in Mexico and the Rocky Mountains.* 1847. Reprint, Glorieta, N.Mex.: Rio Grande Press, 1973.

———. *Life in the Far West.* Edited by LeRoy R. Hafen. 1849. Reprint, Norman: University of Oklahoma Press, 1951.

———. *Ruxton of the Rockies.* Collected by Clyde Porter and Mae Reed Porter. Edited by LeRoy R. Hafen. Norman: University of Oklahoma Press, 1950.

Simmons, Marc, ed. *On the Santa Fe Trail.* Lawrence: University Press of Kansas, 1986. This is a collection of twelve original narratives and reports from the Trail's middle years (1840s–1860s).

Smith, Henry. "Henry Smith's Recollections, 1863." In *On the Santa Fe Trail*, edited by Marc Simmons, 64–72. Lawrence: University Press of Kansas, 1986.

Steck, Michael. "Trail Letter by Dr. Michael Steck, 1852." In *On the Santa Fe Trail*, edited by Marc Simmons, 18–27. Lawrence: University Press of Kansas, 1986.

Storrs, Augustus, and Alphonso Wetmore. *Santa Fé Trail, First Reports: 1825.* 1825; facsimile reprint, Houston, Tex.: Stagecoach Press, 1960.

Vanderwalker, George E. "Reminiscences of George E. Vanderwalker, 1864." In *On the Santa Fe Trail*, edited by Marc Simmons, 84–95. Lawrence: University Press of Kansas, 1986.

Waugh, Alfred S. *Travels in Search of the Elephant: The Wanderings of Alfred S. Waugh, Artist, in Louisiana, Missouri, and Santa Fe, 1845–1846.* Edited and annotated by John Francis McDermott. St. Louis: Missouri Historical Society, 1951.

Webb, James Josiah. *Adventures in the Santa Fé Trade, 1844–1847.* Glendale, Calif.: Arthur H. Clark Co., 1931. Reprint edited by Ralph P. Bieber with an introduction by Mark L. Gardner. Lincoln: University of Nebraska Press, 1995.

———. "On the Trail of My Great-Great Grandfather James Josiah Webb," by Eugenie Webb Maine. *Wagon Tracks* 28, no. 2 (February 2014): 14–15.

———. "The Papers of James J. Webb, Santa Fe Merchant, 1844–1861." Edited by Ralph P. Bieber. *Washington University Studies* 11 (April 1924): 255–305. See also Webb under Kingsbury, John M.

Weichselbaum, Theodore. "Theodore Weichselbaum: Merchant on the Trail." Parts 1 and 2. *Wagon Tracks* 6, no. 1 (November 1991): 9–10; 6, no. 2 (February 1992): 13–15. The first part includes "The Statement of Theodore Weichselbaum."

Wetmore, Alphonso. "Alphonso Wetmore Letters." Edited by Leo E. Oliva. *Wagon Tracks* 14, no. 2 (February 2000): 10.

———. "Alphonso Wetmore's Report of a Journey to Santa Fe in 1828. Edited by Harry C. Myers. *Wagon Tracks* 16, no. 4 (August 2002): 9–17.

———. "Major Alphonso Wetmore's Diary of a Journey to Santa Fe, 1828." *Missouri Historical Review* 8, no. 4 (July 1914): 177–97.

———. *Santa Fé Trail, First Reports: 1825.* 1825; facsimile reprint, Houston, Tex.: Stagecoach Press, 1960.

White, David A., ed. *News of the Plains and Rockies, 1803–1865: Original Narratives of Overland Travels and Adventures Selected from the Wagner-Camp Bibliography of Western America.* Vol. 2, C, "Santa Fe Adventurers, 1818–1843"; Vol. 2, D, "Settlers, 1819–1865." Spokane, Wash.: Arthur H. Clark Co., 1996.

Wislizenus, Frederick A. *A Journey to the Rocky Mountains in the Year 1839.* 1912. Facsimile reprint, Glorieta, N.Mex.: Rio Grande Press, 1969.

The Santa Fe Trail

Histories and Related Subjects

Arnold, Samuel P. *Eating Up the Santa Fe Trail.* Illustrated by Carrie Arnold. Niwot: University Press of Colorado, 1990.

Barry, Louise. *The Beginning of the West: Annals of the Kansas Gateway to the American West, 1540–1854.* Topeka: Kansas State Historical Society, 1972.

Beachum, Larry Mahon. *William Becknell: Father of the Santa Fe Trade.* El Paso: Texas Western Press, 1982.

Boyle, Susan Calafate. *Los Capitalistas: Hispano Merchants and the Santa Fe Trade.* Albuquerque: University of New Mexico Press, 1997.

———. *Comerciantes, Arrieros y Peones: The Hispanos and the Santa Fe Trade.* Professional Paper, no. 54. Santa Fe, N.Mex.: National Park Service, Southwest Regional Office, 1995.

Brown, William E. *The Santa Fe Trail: National Park Service 1963 Historic Sites Survey.* St. Louis, Mo.: Patrice Press, 1988.

Chalfant, William Y. *Dangerous Passage: The Santa Fe Trail and the Mexican War.* Foreword by Marc Simmons. Norman: University of Oklahoma Press, 1994.

Chaput, Donald. *François X. Aubry: Trader, Trailmaker and Voyageur in the Southwest, 1846–1854.* Glendale, Calif.: Arthur H. Clark Co., 1975.

Clapsaddle, David K. *A Directory of Santa Fe Trail Sites, Associated with the Wet and Dry Routes . . . and the Fort Hays–Fort Dodge Road.* Larned, Kan.: Wet/ Dry Routes Chapter of the Santa Fe Trail Association, 1999. This compilation locates the courses of these routes and of the little-known Fort Hays– Fort Dodge Road, which intersected with the Trail.

———. "The Dry Route Revisited." *Overland Journal* (Summer 1999): 2–8. Reprinted in *Wagon Tracks* 14, no. 1 (November 1999): 8–11.

Coleman, Jon T. *Here Lies Hugh Glass: A Mountain Man, A Bear, and the Rise of the American Nation.* American Portrait Series. New York: Hill & Wang, 2012.

Connor, Seymour V., and Jimmy M. Skaggs. *Broadcloth and Britches: The Santa Fe Trade.* College Station: Texas A&M University Press, 1977.

Coyner, David H. *The Lost Trappers.* 1847. Reprint edited by and with an introduction by David J. Weber. Albuquerque: University of New Mexico Press, 1970.

Crutchfield, James A. *The Santa Fe Trail.* Plano: Republic of Texas Press, 1996.

Davis, H. Denny. "Franklin: Cradle of the Trade." *Wagon Tracks* 7, no. 3 (May 1993): 11–17.

Dayton, Duncan. *Lewis and Clark: The Journey of the Corps of Discovery.* New York: Alfred A. Knopf, 1997.

Dodge, Richard Irving. *The Plains of the Great West and Their Inhabitants.* New York: G. P. Putnam's Sons, 1877.

Duffus, Robert L. *The Santa Fe Trail.* New York: Longmans, Green & Co., 1930. Reprints, Albuquerque: University of New Mexico Press, 1972, 1999.

Frazer, Robert W. *Forts of the West.* Norman: University of Oklahoma Press, 1965.

Gardner, Mark L. *Bent's Old Fort National Historic Site.* Tucson, Ariz.: Western National Parks Association, 1998. 16 pp.

———. *Fort Union National Monument.* Tucson, Ariz.: Western National Parks Association, 2005.

———, ed. *The Mexican Road: Trade, Travel, and Confrontation on the Santa Fe Trail.* Manhattan, Kan.: Sunflower University Press, 1989.

———. *Santa Fe Trail: National Historic Trail.* Edited by Ron Foreman. Tucson, Ariz.: Western National Parks Association, 2008. 15 pp.

———. *Wagons for the Santa Fe Trade: Wheeled Vehicles and Their Makers, 1822– 1880.* Albuquerque: University of New Mexico Press, 2000.

Hall, Thomas B., M.D. *Medicine on the Santa Fe Trail.* Arrow Rock, Mo.: Friends of Arrow Rock, 1987.

Hamilton, Jean Tyree. *Arrow Rock: Where Wheels Started West.* Arrow Rock, Mo.: Friends of Arrow Rock, 1972.

Hammond, George P. *The Adventures of Alexander Barclay, Mountain Man.* Denver, Colo.: Old West Publishing Co., 1976.

Hardeman, Nicholas Perkins. *Wilderness Calling.* Knoxville: University of Tennessee Press, 1977.

Holling, Holling Clancy. *Tree in the Trail.* Boston: Houghton Mifflin, 1942. Reprinted numerous times since its first appearance, this illustrated classic has long been cherished by young and older readers alike.

Hollon, W. Eugene. *The Lost Pathfinder: Zebulon Montgomery Pike.* Norman: University of Oklahoma Press, 1949.

Hyslop, Stephen G. *Bound for Santa Fe: The Road to New Mexico and the American Conquest, 1806–1848.* Norman: University of Oklahoma Press, 2002.

Inman, Colonel Henry. *The Old Santa Fé Trail.* Topeka, Kan.: Crane & Co. 1899. Reprint, Williamstown, Mass.: Corner House, 1977.

Lavender, David. *Bent's Fort.* Garden City, N.Y.: Doubleday & Co., 1954. Reprint, Lincoln: University of Nebraska Press, 1979.

Martin, Gene, and Mary Martin. *Trail Dust: A Quick Picture History of the Santa Fe Trail.* Martin Associates, 1991.

Mead, J. R. *Hunting and Trading on the Great Plains, 1859–1875.* Edited by Schuyler Jones. Norman: University of Oklahoma Press, 1986.

Meline, James F. *Two Thousand Miles on Horseback, Santa Fé and Back; A Summer Tour through Kansas, Nebraska, Colorado, and New Mexico in the Year 1866.* Albuquerque, N.Mex.: Horn & Wallace, 1966.

Moorhead, Max L. *New Mexico's Royal Road: Trade and Travel on the Chihuahua Trail.* Norman: University of Oklahoma, 1995. Moorhead could have added at the end of the title "and the Santa Fe Trail," because this book devotes a considerable amount of attention to the Trail up to 1848.

O'Brien, William Patrick. *Merchants of Independence: International Trade on the Santa Fe Trail, 1827–1860.* Kirksville, Mo.: Truman State University Press, 2014.

Oliva, Leo E., ed. *Adventure on the Santa Fe Trail: Selected Papers from the Santa Fe Trail Symposium.* Hutchinson, Kansas, September 24–26, 1987. Topeka: Kansas State Historical Society, 1988.

———, ed. *Confrontation on the Santa Fe Trail: Selected Papers from the Santa Fe Trail Symposia, La Junta, Colorado, 1992, and Great Bend, Kansas, 1995.* Foreword by Dave Webb. Woodston, Kan.: Santa Fe Trail Association Publications, 1996.

———. "The 1829 Escorts." *Confrontation on the Santa Fe Trail.* Woodston, Kan.: Santa Fe Trail Association, 1996.

———. *Fort Atkinson on the Santa Fe Trail, 1850–1854. Kansas Historical Quarterly* (Summer 1974). Reprint, Wet/Dry Routes Chapter, Santa Fe Trail Association, Larned, Kansas, 2007.

———. *Fort Dodge: Sentry of the Western Plains.* Topeka: Kansas State Historical Society, 1998.

————. *Fort Larned: Guardian of the Santa Fé Trail.* Topeka: Kansas State Historical Society, 1997.

————. *Fort Larned on the Santa Fe Trail.* Topeka: Kansas State Historical Society, 1982.

————. *Fort Union and the Frontier Army in the Southwest.* Santa Fe: National Park Service, Division of History, 1993.

————. "'Our Friend Melgares': Spaniards and Mexicans and the Santa Fe Road." *Wagon Tracks* 29, no. 1 (November 2014): 12–18.

————. *Soldiers on the Santa Fe Trail.* Norman: University of Oklahoma Press, 1967.

Olsen, Michael L., and Harry C. Myers. "The Diary of Pedro Ignacio Gallego Wherein 400 Soldiers Following the Trail of Comanches Met William Becknell on His First Trip to Santa Fe." *Wagon Tracks* 7, no. 1 (November 1992): 1, 15–20.

Rittenhouse, Jack D. *Trail of Commerce and Conquest: A Brief History of the Road to Santa Fe.* Larned, Kan.: Santa Fe Trail Association, 2000.

Sandoval, David A. "Gnats, Goods, and Greasers: Mexican Merchants on the Santa Fe Trail." In *The Mexican Road: Trade, Travel, and Confrontation on the Santa Fe Trail,* ed. Mark L. Gardner, 22–31. Manhattan, Kan.: Sunflower University Press, 1989.

Simmons, Marc. *The Old Trail to Santa Fe: Collected Essays.* Albuquerque: University of New Mexico Press, 1996.

————. *Opening the Santa Fe Trail: One Hundred and Fifty Years, 1821–1971.* Cerrillos, N.Mex.: Cerrillos Press, 1971. This is a seven-page keepsake commemorating the 150th anniversary of the old Trail.

————. *The Santa Fe Trail Association: A History of Its First Decade, 1986–1996.* Larned, Kan.: Santa Fe Trail Assocation, 1997. 34 pp.

————. "The Santa Fe Trail: Highway of Commerce." In *Trails West,* 8–39. Washington, D.C.: National Geographic Society, 1979.

————. *Yesterday in Santa Fe: Episodes in a Turbulent History.* Revised ed. Santa Fe, N.Mex.: Sunstone Press, 1987.

Slusher, Roger. "Lexington and the Santa Fe Trail." *Wagon Tracks* 5, no. 4 (August 1991): 6–9.

Strate, David K. *Sentinel to the Cimarron: The Frontier Experience of Fort Dodge, Kansas.* Dodge City: Cultural Heritage and Arts Center, 1970.

Taylor, Morris. *First Mail West: Stagecoach Lines on the Santa Fe Trail.* Albuquerque: University of New Mexico Press, 1971. Reprint, 2000.

Thompson, Gerald. *Edward F. Beale and the American West.* Albuquerque: University of New Mexico Press, 1983.

Utley, Robert M. *Fort Union and the Santa Fe Trail.* El Paso: Texas Western Press, University of Texas at El Paso, 1989.

Vestal, Stanley (pseud. of Walter Stanley Campbell). *The Old Santa Fe Trail*. 1939. Reprinted with an introduction by Marc Simmons to the Bison Books edition. Lincoln: University of Nebraska Press, 1996.

Walker, Henry P. *The Wagonmasters*. Norman: University of Oklahoma Press, 1966.

Webb, Dave. *Fort Larned Adventures: An Activity Book*. Protection, Kan.: Comanche Press, 1997. Illustrated by Phillip R. Buntin. Leo E. Oliva and George Elmore, historical consultants. 96 pp.

———. *Santa Fe Trail Adventures: An Activity Book*. Dodge City: Kansas Heritage Center, 1999. Illustrated by Phillip R. Buntin. Leo E. Oliva, historical consultant. 88 pp. This is an expanded edition of his earlier *Adventures with the Santa Fe Trail* and is of interest to anyone learning about the Trail.

Weber, David J., ed. and trans. *The Extranjeros: Selected Documents from the Mexican Side of the Santa Fe Trail, 1825–1844*. Santa Fe, N.Mex.: Stagecoach Press, 1967.

Wetzel, David N., ed. *The Santa Fe Trail: New Perspectives*. Essays and Monographs in Colorado History, no. 6. Denver: Colorado Historical Society, 1987. This is a special issue consisting of papers from the first Santa Fe Trail Symposium held in Trinidad, Colorado, September 12–14, 1986.

Woodridge, Rhoda. *Fort Osage*. Independence, Mo.: Independence Press, 1983.

Young, Otis E. *The First Military Escort on the Santa Fe Trail, 1829; From the Journal and Reports of Major Bennet Riley and Lieutenant Philip St. George Cooke*. Glendale, Calif.: Arthur H. Clark Co., 1952.

Trail Guides

Carter, Anne, and David Carter. *Mulberries and Prickly Pear*. Independence, Mo.: Arrow Press, 1991. Ancestors of the authors traveled the Santa Fe Trail. The Carters rode their horses on parts of the old Trail and on modern roads.

Cordes, Kathleen Ann. *America's National Historic Trails*. Norman: University of Oklahoma Press, 1999.

Curtiss, Frank. *Re-riding History: Horseback Over the Santa Fe Trail*. Santa Fe, N.Mex.: Sunstone Press, 1997.

Franzwa, Gregory M. *Maps of the Santa Fe Trail*. St. Louis, Mo.: Patrice Press, 1989. An eighteen-page booklet of errata sheets providing maps and annotations was published in 1990 to accompany this book.

———. *The Santa Fe Trail Revisited*. St. Louis: Patrice Press, 1989.

Hill, William E. *The Santa Fe Trail, Yesterday and Today*. Caldwell, Idaho: Caxton Printers, 1992.

Martin, Gene, and Mary Martin. *Trail Dust*. Boulder, Colo.: Johnson Books, 1972.

Penner, Marci. *Eight Wonders of Kansas Guidebook*. Inman, Kan.: Kansas Sampler Foundation, 2011. Historic Council Grove is one of the eight wonders of

Kansas history. A number of other Santa Fe Trail sites are among the 216 finalists in the wonders categories. Includes hundreds of color photographs and interesting details.

Pinkerton, Elaine. *The Santa Fe Trail by Bicycle.* Santa Fe, N.Mex.: Red Crane Books, 1993.

Ross, Inez. *Without a Wagon on the Santa Fe Trail: Hiking into History.* 2nd ed. Los Alamos, N.Mex.: Ashley House, 2004.

Simmons, Marc, and Hal Jackson. *Following the Santa Fe Trail: A Guide for Modern Travelers.* 3rd ed., revised and expanded. Santa Fe, N.Mex.: Ancient City Press, 2001.

Stocking, Hobart E. *The Road to Santa Fe.* New York: Hastings House, 1971.

White, William W. *The Santa Fe Trail by Air: A Pilot's Guide to the Santa Fe Trail.* Logan, Utah: Western Airtrails, 1996.

———. "Revisiting the Santa Fe Trail." *SWAviator* (April/May 2001): 18–23, 31. With color photos and maps.

Photography and Art

Baca, Elmo. "Hitting the Trail: The Long, Winding Route to Santa Fe." *New Mexico Magazine* 75, no. 2 (February 1997): 42–51. Color photography by Paul Logsdon and Mark Nohl.

Chávez, Fray Angélico. "A Century and a Half Old: Ruts of the Santa Fe Trail. *New Mexico Magazine* (Fiesta Issue) 50, nos. 7–8 (1972): 18–29. Color photography by C. M. Montgomery. This outstanding article is among the longest ever published in this long-running magazine.

Dulle, Ronald J. *Tracing the Santa Fe Trail: Today's Views and Yesterday's Voices.* Missoula, Mont.: Mountain Press Publishing Co., 2011. Excellent color photography by Ronald J. Dulle.

Findley, Rowe. "Along the Santa Fe Trail." Photography by Bruce Dale. *National Geographic,* March 1991, 98–123.

Gardner, Mark Lee. "Romancing the Trail: A 175-Year Love Affair." Story and color photography by Mark Lee Gardner. *New Mexico Magazine* 75, no. 2 (February 1997): 52–59.

———. "Harbinger of Change: The Santa Fe Trail." In "American Cowboy Legends, Collector's Edition." A special edition on the Santa Fe Trail. *American Cowboy* (2013): 32–37. Illustrations by Jim Carson and photography by Mark Lee Gardner.

Kil, Ronald, and the Brownell Museum of the Southwest. *A Pictorial History of the Southwest and the Santa Fe Trail: Featuring the Artwork and Narrative of Ronald Kil.* Raton, N.Mex.: Frank Brownell Museum of the Southwest and the Bud and Willa Eyman Library, [2010].

Bibliography

Murphy, Dan, and Bruce Hucko. *Santa Fe Trail, Voyage of Discovery: The Story behind the Scenery.* Las Vegas, Nev.: KC Publications, 1998. Excellent color photography throughout with accompanying story and numerous quotes from writings about the Trail and travel on it beginning with William Becknell in 1821.

Parkison, Jami. *Path to Glory: A Pictorial Celebration of the Santa Fe Trail.* Kansas City, Mo.: Highwater Editions, 1996. Published to commemorate the Trail's 175th anniversary, this book includes an overview of Trail history with an excellent selection of illustrations.

Simmons, Marc, and Joan Myers. *Along the Santa Fe Trail.* Essay by Marc Simmons; photography by Joan Myers. Albuquerque: University of New Mexico Press, 1986.

Websites

Santa Fe Trail Association (official website), http://www.santafetrail.org
Santa Fe Trail Center, Larned, Kansas, http://www.santafetrailcenter.org
Santa Fe National Historic Trail, http://www.nps.gov.safe/index.htm
Bent's Old Fort National Historic Site, Colorado, http://www.nps.gov/beol/index.htm
Fort Larned, Kansas, http://www.nps.gov/fols/index.htm
Fort Union National Monument, New Mexico, http://www.nps.gov/foun/index.htm
National Frontier Trails Museum, Missouri, http://www.ci.independence.mo.us/nftm
Pathways across America (information on trails), http://www.pnts.org/pathways

The Animals

Grizzlies and Black Bears

Brown, David E. *The Grizzly in the Southwest: Documentary of an Extinction.* Foreword by Frank C. Craighead, Jr. Norman: University of Oklahoma Press, 1985. Reprint, 1996.

Busch, Robert H. *The Grizzly Almanac.* New York: Lyons Press, 2000.

——. *Valley of the Grizzlies.* New York: St. Martin's Press, 1998.

Craighead, Lance. *Bears of the World.* Stillwater, Minn.: Voyageur Press, 2003.

Manfred, Fredrick F. *Lord Grizzly.* New York: McGraw-Hill, 1954.

McNamee, Thomas. *The Grizzly Bear.* New York: Alfred A. Knopf, 1984.

Powell, R. A., J. W. Zimmerman, and D. E. Seaman. *Ecology and Behaviour of North American Black Bears: Home Range, Habitat, and Social Organization.* New York: Chapman & Hall, 1997.

Bibliography

Buffalo

Dary, David. *The Buffalo Book.* Chicago: Swallow Press, 1974.

Gard, Wayne. *The Great Buffalo Hunt.* New York: Alfred A. Knopf, 1959. Reprint, Lincoln: University of Nebraska Press, 1968.

Geist, Valerius. *Buffalo Nation: History and Legend of the North American Bison.* Stillwater, Minn.: Voyageur Press, 1996.

Lott, Dale E. *American Bison: A Natural History.* Berkeley: University of California Press, 2002.

McHugh, Tom. *The Time of the Buffalo.* New York: Alfred A. Knopf, 1972.

Punke, Michael. *Last Stand: George Bird Grinnell, the Battle to Save the Buffalo, and the Birth of the American West.* New York: Smithsonian Books/Collins, 2007.

Rudner, Ruth. *A Chorus of Buffalo: A Personal Portrait of an American Icon.* New York: Marlowe & Co., 2004.

Simmons, Marc. "On the Buffalo Trail." Trail Dust (column). *Santa Fe Reporter,* February 3–9, 1993, 24.

Burros (Donkeys)

Brookshier, Frank. *The Burro.* Norman: University of Oklahoma Press, 1974.

McKnight, T. L. "The Feral Burro in the United States: Distribution and Problems." *Journal of Wildlife Management* 22 (1958): 163–78.

Simmons, Marc. "Mexican Burro Representative of Southwest Trails." Trail Dust (column). *Santa Fe New Mexican,* March 10, 2001, B1, B3.

———. "The Romance of the Burro." Trail Dust. *Santa Fe Reporter,* July 19–25, 1995, 27.

Vilda, Les. "A Modern Encounter with the Trail." *Wagon Tracks* 3, no. 3 (May 1989): 6.

Woodward, Susan L. "The Living Relatives of the Horse." Chapter 10 in *Horses through Time,* edited by Sandra L. Olsen. Boulder, Colo.: Roberts Rinehart Publishers for Carnegie Museum of Natural History, 1995. In particular, see pages 202–208 about donkeys, burros, and mules.

Coyotes

Dobie, J. Frank. *The Voice of the Coyote.* Boston: Little, Brown, and Co., 1950. Reprint, Lincoln: University of Nebraska Press, 2006.

Fox, Camilla H., and Christopher M. Papouchis. *Coyotes in Our Midst: Coexisting with an Adaptable and Resilient Carnivore.* Sacramento, Calif.: Animal Protection Institute, 2005.

Gard, Wayne. *The World of the Coyote.* San Francisco: Sierra Club Books, 1994.

McClennen, N., R. Wigglesworth, and S. H. Anderson. *The Effect of Suburban and Agricultural Development on the Activity Pattern of Coyotes (Canis latrans). American Midland Naturalist* 146, 27–36.

Meinzer, Wyman. *Coyote*. Lubbock: Texas Tech University Press, 1995.

Simmons, Marc. "Coyote's Life Is Tough." Trail Dust (column). *Santa Fe Reporter,* January 19, 1983, 17.

Young, Stanley P. *The Clever Coyote; Part I: Its History, Life Habits, Economic Status, and Control.* Harrisburg, Pa., Stackpole Co., Washington, D.C.: Wildlife Management Institute, 1951. Reprint, Lincoln: University of Nebraska Press, 1978.

Dogs

Clutton-Brock, Juliet. "Origins of the Dog: Domestication and Early History." In *The Domestic Dog: Its Evolution, Behaviour, and Interactions with People,* ed. James Serpell. Cambridge: Cambridge University Press, 1995.

Coppinger, Raymond, and Lorna Coppinger. *Dogs: A Startling New Understanding of Canine Origin, Behavior, and Evolution.* New York: Scribner, 2001; Chicago: University of Chicago Press, 2002.

Derr, Mark. *A Dog's History of America: How Our Best Friend Explored, Conquered, and Settled a Continent.* New York: North Point Press, 2004.

———. *Dog's Best Friend: Annals of the Dog-Human Relationship.* New York: Henry Holt, 1997.

———. *How the Dog Became the Dog: From Wolves to Our Best Friends.* New York: Overlook Press, Peter Mayer, Publishers, 2011.

Pferd, William, III. *Dogs of the American Indians.* Edited by William W. Denlinger and R. Annabel Rathman. Fairfax, Va.: Denlinger's, 1987.

Schwartz, Marion. *A History of Dogs in the Early Americas.* New Haven, Conn.: Yale University Press, 1997.

Simmons, Marc. "New Mexico Historians Have Forgotten the Role of the Dog." History (column). *New Mexico Independent* (Albuquerque), December 1, 1978, 9.

———. "Sheep Dogs in the Southwest." Trail Dust (column). *Santa Fe Reporter,* July 11–17, 1990, 15.

Horses (Domestic)

See also Mustangs (Wild Horses).

Clark, LaVerne H. *They Sang for Horses: The Impact of the Horse on Navajo and Apache Folklore.* Niwot: University Press of Colorado, 2001.

Clutton-Brock, Juliet. *Horse Power: A History of the Horse and the Donkey in Human Societies.* Cambridge, Mass.: Harvard University Press, 1992.

Culley, John H. *Cattle, Horses, and Men of the Western Range.* Tucson: University of Arizona Press, 1984.

Ensminger, M. E. *Horses and Horsemanship: Animal Agriculture Series.* 6th ed. Danville, Ill.: Interstate Publishing, 1990.

Olsen, Sandra L., ed. *Horses through Time.* Boulder, Colo.: Roberts Rinehart Publishers for Carnegie Museum of Natural History, 1995. Includes horses' relatives around the world.

Mules

American Donkey and Mule Society, http://www.lovelongears.com.

Attar, Cynthia. *The Mule Companion: A Guide to Understanding the Mule.* Portland, Ore.: Partner Communications, 1998.

Ewing, Floyd F., Jr. "The Mule as a Factor in the Development of the Southwest." *Arizona and the West* 5, no. 4 (Winter 1963): 315–20.

Simmons, Marc. "*Arrieria:* The Art of Mexican Muleteering." *Mr. Longears* (journal of the American Donkey and Mule Society) 7, no. 1 (Spring 1975): 31–33. Reprint, *Southwest Heritage* 7, no. 1 (Spring 1977): 2–5. Reprint, *Chronicle of the Horse* 40, no. 16 (April 22, 1977): 28–29.

———. "The Mule: An Unsung Hero of the Desert Southwest." Trail Dust (column). *Santa Fe New Mexican,* November 15, 2003, B1.

———. "Sharing the Load: The Army Mule." Trail Dust (column). *Santa Fe Reporter,* September 1–7, 1993, 21.

Mustangs (Wild Horses)

Dobie, J. Frank. *The Mustangs.* Illustrated by Charles B. Wilson. New York: Bramhall House, 1952; Boston: Little, Brown, 1952.

Flores, Dan. *Unbroken Spirit: The Wild Horse in the American Landscape.* Edited by Frances B. Clymer and Charles R. Preston. Cody, Wyo.: Buffalo Bill Historical Center, 1999.

Green, Ben K. *A Thousand Miles of Mustangin'.* Silver City, N.Mex.: High-Lonesome Books, 1999.

Hockensmith, John S. *Spanish Mustangs in the Great American West: Return of the Horse.* Editor and co-writer Michele MacDonald. Norman: University of Oklahoma Press, 2009.

Ryden, Hope. *America's Last Wild Horses.* New York: Lyons & Burford, 2005.

———. *Wild Horses I Have Known.* New York: Clarion Books, 1999.

Silverstein, Alvin, and Virginia Silverstein. *The Mustang.* Brookfield, Conn.: Millbrook Press, 1997.

Simmons, Marc. *Mustang Days.* Trail Dust (column). *Santa Fe Reporter,* November 8–14, 1995, 29.

Stillman, Deanne. *Mustang: The Saga of the Wild Horse in the American West.* Boston: Houghton Mifflin, 2008.

Bibliography

Oxen

Clapsaddle, David K. "Old Dan and His Traveling Companions: Oxen on the Santa Fe Trail." *Wagon Tracks* 22, no. 2 (February 2008): 10.

Conroy, Drew. *Oxen, A Teamster's Guide.* 2nd ed. North Adams, Mass.: Storey, 2007.

Conroy, Drew, and Dwight Barney. *The Oxen Handbook.* Marysville, Mo.: D. Butler, 1986.

Pelzer, Louis. *The Cattlemen's Frontier: A Record of the Trans-Mississippi Cattle Industry from Oxen Trains to Pooling Companies, 1850–1890.* Glendale, Calif.: Arthur H. Clark Co., 1936.

Rouse, John E. *The Criollo Spanish Cattle of the Americas.* Norman: University of Oklahoma Press, 1977.

Simmons, Marc, "Little-known Ox Lore Gives Animals Their Due," Trail Dust (column). *Santa Fe New Mexican,* August 27, 2005, C3.

———. "Oxen versus Mules." In *The Old Trail to Santa Fe: Collected Essays,* 143–46. Albuquerque: University of New Mexico Press, 1996.

Prairie Chicken

Cable, Ted T., et al. *Birds of the Cimarron National Grassland.* Technical Report RM-GTR-28. Fort Collins, Colo.: U.S. Department of Agriculture, Forest Service, 1996.

Church, Kevin. "A Rite of Courtship." *Kansas Wildlife and Parks,* March/April 1989: 2.

Crawford, John A. *A Bibliography of the Lesser Prairie Chicken, 1873–1980.* Fort Collins, Colo.: Rocky Mountain Forest and Range Experiment Station, Forest Service, U.S. Dept. of Agriculture, November 1980.

Horak, Gerald J. "The Prairie Bird." *Kansas Wildlife* 44, no. 8 (November/December 1987): 8–12.

Manes, Rob. "Phantoms of the Prairie." *Kansas Wildlife* 40, no. 8 (November/December 1983): 19.

Peterson, Roger Tory. *A Field Guide to Western Birds.* Boston: Houghton Mifflin, 1990.

Sibley, David. *The Sibley Guide to Birds.* New York: Alfred A. Knopf, 2000.

Sibley, David, et al., *The Sibley Guide to Bird Life and Behavior.* New York: Alfred A. Knopf, 2001.

Zimmer, Kevin J. *The Western Bird Watcher.* Englewood Cliffs, N.J.: Prentice Hall, 1985.

Prairie Dogs

Graves, Russell A. *The Prairie Dog: Sentinel of the Plains.* Lubbock: Texas Tech University Press, 2001.

Bibliography

Hoogland, John L. *The Black-Tailed Prairie Dog: Social Life of a Burrowing Mammal*. Chicago: University of Chicago Press, 1995.

———. *Conservation of the Black-Tailed Prairie Dog: Saving North America's Western Grasslands*. Washington, D.C.: Island Press, 2006.

Johnsgard, Paul A. *Prairie Dog Empire: A Saga of the Shortgrass Prairie*. Lincoln: University of Nebraska Press, 2005.

Kotliar, N. B., et al. "A Critical Review of Assumptions about the Prairie Dog as a Keystone Species." *Environmental Management* 24, no. 2 (1999): 177–92.

Long, Michael E. "The Vanishing Prairie Dog." *National Geographic* 193, no. 4 (April 1998): 118–31.

Patent, Dorothy Hinshaw. *Prairie Dogs*. New York: Clarion, 1993.

Pronghorn

Biennial Pronghorn Workshop 2014. Pronghorn Population Count in the United States and Canada, 2013, Sul Ross State University, Alpine, Texas, May 12–14, 2012.

Bromley, Peter T. *Aspects of the Behavioural Ecology and Sociobiology of the Pronghorn (Antilocapra americana)*. Doctoral diss., University of Calgary, Calgary, Alberta, Canada. 1977.

Byers, John A. *American Pronghorn: Social Adaptations and the Ghost of Predators Past*. Chicago: University of Chicago Press, 1997.

Geist, Valerius. *Antelope Country; Pronghorns: The Last Americans*. Iola, Wisc.: Krause Publications, 2001.

Schaefer, John. "An American Loner." In *An American Bestiary*, 18–26. Foreword by James S. Findley. Boston: Houghton Mifflin, 1975.

Turbak, Gary. *Pronghorn: Portrait of the American Antelope*. Flagstaff, Ariz.: Northland Press, 1995.

Rattlesnakes

Dobie, J. Frank. *Rattlesnakes*. Boston: Little, Brown, 1965.

Hubbs, Brian, and Brendan O'Connor. *A Guide to the Rattlesnake and Other Venomous Serpents of the United States*. Tempe, Ariz.: Tricolor Books, 2012.

Klauber, Laurence. *Rattlesnakes: Their Habits, Life Histories and Influence on Mankind*. Berkeley: University of California Press, 1982.

Mattison, Chris. *Rattler: A Natural History of Rattlesnakes*. London: Blanford; New York: Sterling, 1996.

Myers, Bob. "How Dangerous Are Rattlesnakes?" Pamphlet. Albuquerque, N.Mex.: American International Rattlesnake Museum, 1991. Bob Myers is the director of this fascinating museum in Albuquerque's Old Town.

Bibliography

Palmer, Thomas. *Landscape with Reptile: Rattlesnakes in an Urban World*. New York: Ticknor & Fields, 1992.

Rubio, Manny. *Rattlesnake: Portrait of a Predator*. Washington, D.C.: Smithsonian Books, 1998.

Simmons, Marc. "Some Rattlesnake Lore." Trail Dust (column). *Santa Fe Reporter*, November 24–30, 1993, 16.

Roadrunners

Dobie, J. Frank. "The Roadrunner in Fact and Folk-lore." *Texas Ornithological Society Newsletter* 4, no. 4 (May 1, 1956): 1.

———, Mody C. Bratwright, and Harry H. Ransom, eds. *In the Shadow of History*. Texas Folklore Society, publication no. 15. 1939. Reprint, Dallas: Southern Methodist University Press, 1980.

Meinzer, Wyman. *The Roadrunner*. Lubbock: Texas Tech University Press, 1993. This book and Meinzer's *Coyote* have outstanding color photographs.

Simmons, Marc. "Roadrunner Lore." New Mexico Scrapbook (column). *Prime Time* (Albuquerque), November 24–30, 1993, 16.

Whitson, Martha A. "The Roadrunner—Clown of the Desert." Photographs by Bruce Dale. *National Geographic* (May 1983): 694–702.

Wolves

Boitani, Luigi, Paul C. Paquet, and Marco Musiani, eds. *A New Era for Wolves and People: Wolf Recovery, Human Attitudes, and Policy*. Calgary, Alberta, Canada: University of Calgary Press, 2009.

Brown, David E. *The Wolf in the Southwest: The Making of an Endangered Species*. Silver City, N.Mex.: High-Lonesome Books, 2002.

Busch, Robert H. *The Wolf Almanac*. Foreword by Rick Bass. New York: Lyons & Burford, 1995. Includes evolution, distribution, physiology, and behavior. It also has an excellent bibliography.

Holaday, Bobbie. *Return of the Mexican Gray Wolf: Back to the Blue*. Foreword by L. David Mech. Tucson: University of Arizona Press, 2003.

Mech, L. David. *The Wolf: The Ecology and Behavior of an Endangered Species*. New York: Natural History Press, Doubleday, 1970.

Musiani, Marco, Luigi Boitani, and Paul C. Paquet, eds. *The World of Wolves: New Perspectives on Ecology, Behaviour, and Management*. Calgary, Alberta: University of Calgary Press, 2010.

Peck, Robert Morris. *The Wolf Hunters: A Story of the Buffalo Plains*. Edited and arranged from the manuscript account of Robert M. Peck by George Bird Grinnell. New York: Scribner's Sons, 1914. This historical novel is based on

Bibliography

Peck's wolf hunting and related experiences when he was a soldier stationed at Fort Larned in Kansas during the winter of 1861–62. Brief accounts were also published in the *National Tribune* in 1901.

Simmons, Marc. "Wolves: Symbol of the Old West." Trail Dust (column). *Santa Fe Reporter,* January 14–20, 1998, 28.

Young, Stanley P. *The Wolves of North America. Part 1: History, Life Habits, Economic Status, and Control.* Washington, D.C.: American Wildlife Institute, 1944.

Other Works about Animals

Clutton-Brock, Juliet. *A Natural History of Domesticated Mammals.* Cambridge: Cambridge University Press, 1999.

Cockrum, E. Lendell. *Mammals of the Southwest.* Tucson: University of Arizona Press, 1982.

Feldhamer, George A., Bruce C. Thompson, and Joseph A. Chapman. *Wild Mammals of North America: Biology, Management, and Conservation.* Baltimore: Johns Hopkins University Press, 2003.

Fleharty, Eugene D. *Wild Animals and Settlers on the Great Plains.* Norman: University of Oklahoma Press, 1995.

Kays, Roland W., and Don E. Wilson. *Mammals of North America.* Princeton Field Guides. Princeton, N.J.: Princeton University Press, 2002.

Krunk, Hans. *Hunter and Hunted: Relationships between Carnivores and People.* Cambridge: Cambridge University Press, 2002.

Osborne, Charles, ed. *Domestic Descendants.* New York: Time-Life Films, 1979.

Rudner, Ruth. *Ask Now the Beasts: Our Kinship with Animals Wild and Domestic.* New York: Marlowe & Co., 2006.

Schaefer, Jack. *An American Bestiary.* Foreword by James E. Findley. Boston: Houghton Mifflin, 1975.

Scully, Matthew. *Dominion: The Power of Man, the Suffering of Animals, and the Call to Mercy.* New York: St. Martin's Griffin, 2003.

Seton, Ernest Thompson. *Wild Animals I Have Known.* New York: Charles Scribner's Sons, 1898. Unabridged facsimile reprint, Minneola, N.Y.: Dover, 2000. See "Lobo: The King of Currumpaw," 15–44, and "The Pacing Mustang," 183–222.

Wolff, Jerry, and Paul W. Sherman. *Rodent Societies: An Ecological and Evolutionary Perspective.* Chicago: University of Chicago Press, 2007.

Index

Index

Lott, Dale F., *continued*
   See also Dodge, Colonel Richard I.;
   Seton, Ernest Thompson; Shaw, Jim
Louisiana Purchase Territory, 8, 36, 90
Lower Cimarron Spring (Kans.), 18, 45
Lummis, Charles Fletcher: on burros in
   Santa Fe, 145; survived his "tramp across
   the continent" (Ohio to Los Angeles,
   1885), 145

Magoffin, Samuel, 11, 19; size of Magoffin
   caravan in 1846, 157
Magoffin, Susan Shelby (wife of Samuel),
   11, 21; on antelope and shooting of by
   Samuel, 19–20; on buffalo, 11–12, 158;
   on dog ring, 157–58; mules and oxen on
   Trail, 135; prairie dogs' behavior and
   towns, 33; on profanity, 123, 135; on
   rattlesnakes, 78; on scampers
   (stampedes), 135; on wolves and
   mosquitoes, 157–58
magpies, 35
Majors, Alexander: on freighting, 79, 126;
   on misconception that prairie dogs,
   rattlesnakes, and owls lived together, 83,
   124, 126; on rattlesnakes biting draft
   animals, 79; spoke at Old Plainsmen's
   gatherings, 79, 83, 126. *See also* Old
   Plainsmen's Association; Russell, Majors
   and Waddell (freighting company)
Maltese, Michael, 63
Marmaduke, Meredith Miles, 131
Marmots, 27, 31, 45
Maxwell, Lucien Bonaparte: holder of
   Maxwell Land Grant, 38; owner of
   Rayado Ranch, 111
McClure, James, chased by wolves, 47
McGuillicuddy, Mike, died of rabies at Fort
   Union, 49
McNees Crossing (N.Mex.), named for
   Robert McNees after death in Indian
   attack, 20, 132
Meinzer, Wyman, on roadrunners and
   their sounds, 64
Mescalero Apaches, 45
Mexican donkeys. *See* burros
Mexican gray wolves (*Canis lupus baileyi*).
   See wolves
Mexicans, 55, 61, 82, 109, 118, 122, 130;
   with caravans or trappers, 136, 157, 160;

as lassoing experts, 105. See also
   arrieros; *vaqueros*
Mexican War (1846–48), 6, 9, 89, 105
Mexico, 24, 40, 50, 53, 60, 106, 107;
   boundary with United States at
   Arkansas River, 119; independence from
   Spain, 130
Mexico City, 55, 130
mice, 57, 62, 165
Middle Cimarron Spring (Kans.), 18, 82
Milnesand (N.Mex.), 73
Minnesota, 21, 37, 47, 67
Mississippi River, 17, 35, 92
Missouri, 3, 16, 47, 65, 75, 95; home of
   famous Missouri mule, 131, 138;
   Howard County, 130, 131; leading mule-
   producing state, 118; Saline County, 7,
   149; statehood in 1821, 130. *See also*
   Franklin, Mo. (Old Franklin); New
   Franklin, Mo.
Missouri Mounted Volunteers, in Mexican
   War, 6, 9, 32, 47, 76, 82
Missouri River, 35, 94, 95, 108, 109; bears
   encountered along, 91, 98; Corps of
   Discovery on, 91; freight on Trail
   delivered by steamboats on, 138
Monroe, Daniel, 132
Montana, 12, 24, 114
Moore, John W., 122; became mayor of
   Kansas City, 79; saw immense lair of
   rattlesnakes, 78–79
Moorhead, Max L., on contributions of
   mules to Southwest history, 129
mosquitoes, 27, 78, 158
mountain men, 7, 31, 38, 50, 130. *See also*
   trappers
Mountain Route, xv; Bent's Fort on, 7, 59,
   109; buffalo herds, 5, 14; grizzlies and
   black bears on, 93; mustangs in
   abundance, 109; Purgatoire River near,
   143; Spanish Peaks seen from, 21; toll
   road over rough pass on, 48, 93;
   travelers on, 19, 21, 59, 90, 104, 111. *See
   also* Raton Pass (Colo., N.Mex.); Vilda,
   Les; Wootton, Richens Lacy "Uncle Dick"
mules (hybrid, *Equus caballus x Equus
   assinus, Equus mulis*): advantages of
   using mules over horses or oxen, 132–
   33, 134; bitten by rattlesnakes, 77, 78,
   79–80; contributions to development